Lecture Notes in Physics

Volume 853

For further volumes:
http://www.springer.com/series/5304

The Lecture Notes in Physics

The series Lecture Notes in Physics (LNP), founded in 1969, reports new developments in physics research and teaching—quickly and informally, but with a high quality and the explicit aim to summarize and communicate current knowledge in an accessible way. Books published in this series are conceived as bridging material between advanced graduate textbooks and the forefront of research and to serve three purposes:

- to be a compact and modern up-to-date source of reference on a well-defined topic
- to serve as an accessible introduction to the field to postgraduate students and nonspecialist researchers from related areas
- to be a source of advanced teaching material for specialized seminars, courses and schools

Both monographs and multi-author volumes will be considered for publication. Edited volumes should, however, consist of a very limited number of contributions only. Proceedings will not be considered for LNP.

Volumes published in LNP are disseminated both in print and in electronic formats, the electronic archive being available at springerlink.com. The series content is indexed, abstracted and referenced by many abstracting and information services, bibliographic networks, subscription agencies, library networks, and consortia.

Proposals should be sent to a member of the Editorial Board, or directly to the managing editor at Springer:

Christian Caron
Springer Heidelberg
Physics Editorial Department I
Tiergartenstrasse 17
69121 Heidelberg/Germany
christian.caron@springer.com

Malte Henkel • Dragi Karevski

Editors

Conformal Invariance: an Introduction to Loops, Interfaces and Stochastic Loewner Evolution

 Springer

Editors

Malte Henkel
Dépt. de Physique de la Matière
  et des Matériaux
Institut Jean Lamour (CNRS UMR 7198)
Université de Lorraine Nancy
Vandœuvre-lès-Nancy Cedex
France

Dragi Karevski
Dépt. de Physique de la Matière
  et des Matériaux
Institut Jean Lamour (CNRS UMR 7198)
Université de Lorraine Nancy
Vandœuvre-lès-Nancy Cedex
France

ISSN 0075-8450   e-ISSN 1616-6361
Lecture Notes in Physics
ISBN 978-3-642-27933-1   e-ISBN 978-3-642-27934-8
DOI 10.1007/978-3-642-27934-8
Springer Heidelberg Dordrecht London New York

Library of Congress Control Number: 2012935754

Springer is part of Springer Science+Business Media (www.springer.com)

Pour nos enfants:

Klara,
Léna, Ulysse, Ana-Fleur et
Alexandre Bonaventure

Preface

"*Bien des choses ne sont impossibles que parce qu'on s'est accoutumé à les regarder comme telles.*"

Charles Pinot Duclos (1751)

This volume was grown from lectures given at the atelier "*Modern applications of conformal invariance/Applications modernes de l'invariance conforme*" held in Nancy in march 2011. The enormous progress made possible by the systematic utilisation of methods of conformal field-theory for the precise understanding of critical phenomena, at least in two spatial dimensions, has been understood since almost three decades. Still, much of the impressive success of conformal invariance is related to *local* observables, their exponents and their correlators. Much recent work has been devoted to attempts to understand better the behaviour of *extended* objects, such as interfaces. This volume proposes an informal introduction to current results, methods and open questions, which we hope to be accessible to graduate students and researchers from other fields. Familiarity of the reader with the basic techniques in equilibrium statistical mechanics and critical phenomena is assumed.

In order to make this volume as self-contained as possible, we begin in Chap. 1 with a short and compact introduction to the main concepts and methods of $2D$ conformal invariance. We shall describe therein the main properties of *local* objects, such as primary scaling operators, the energy-momentum tensor, the Virasoro algebra, discuss unitary minimal models and how the partition function can be decomposed into minimal characters, both for bulk critical systems and for critical phenomena near a surface. Besides frequent references to spin systems known from statistical physics, such as the Ising and Potts models, we shall use the conformal field-theory of the free boson as the main paradigmatic example. Besides giving in this way a brief review of the textbook knowledge of those elements of conformal invariance which will be needed in the later chapters, we also introduce the notation to be used throughout this volume. Chapter 2, written by M. Bauer, gives an introduction to the physical and mathematical techniques required to describe an important class of growth models whose behaviour can be studied in great depth—conformally invariant interfaces governed by *Stochastic Loewner Evolution* (SLE). It turned out that the methods required for the study of SLE are quite distinct from those developed previously for the analysis of local observables within conformal field-theory, but they provide a very different and new point of view for the analysis of the behaviour of extended objects, and continue to fascinate physicists and mathematicians alike. Two of the most important properties of SLE, namely conformal invariance and the Domain Markov Property, are carefully explained and the chapter closes with a detailed discussion of the SLE-CFT correspondence. In carrying out this mathematically oriented analysis, a couple of technical assumptions had to be made. Although these may appear to be plausible, it is essential to verify whether these assumptions are actually realised in physically relevant systems. Chapter 3, written by C. Chatelain, presents a review of numerical tests of the basics of the SLE description of interfaces in critical systems. First discussing the most simple spin systems in the Ising and Potts universality classes, the second part of this review explores new ground in investigating to what extent SLE might become applicable in situations where quenched *disorder* becomes relevant. Finally Chap. 4, written by J.L. Jacobsen, addresses the study of two-dimensional loop models and their bulk and surface critical behaviour, analysed with the help of conformal invariance. After a detailed survey of the required graph-theoretical tools, it is shown how to relate the specific examples of the Potts- and $O(n)$-vector models in terms of clusters and oriented loops. In this way the defining model parameters, namely the number of states, can be analytically continued to arbitrary values. These analytic continuations are essential for application of the results to percolation and polymers. Since oriented loops act as level lines of height models, a treatment of these height models, via a geometric Coulomb gas construction, yields the bulk and surface critical exponents. From the underlying Temperley-Lieb algebra, the correct partition function in the continuum-limit can be found, which in turn is needed for the derivation for crossing formulæ in percolation. Novel extensions of the algebraic machinery of the Temperley-Lieb algebra appropriate for boundary conformal field-theory are explained.

Nous remercions chaleureusement les auteurs pour leur grand effort et leur temps devoué à l'écriture de ce volume. MH thanks the organisers of the Programme 'Advanced Conformal Field Theory and Applications' at the Institut Henri Poincaré in Paris for warm hospitality, where the editing was finished. It is a pleasure to thank Springer Verlag and especially C. Caron for having made the writing/editing process as *agréable* as possible.

La science vivante étant dans un processus de renouvellement permanent et fructueux, nous espérons évidemment que les générations futures exploreront les cieux qui pour nous seront restés des *mundi incogniti*. Nous dédions ce volume à nos enfants: à Klara, et à Léna, Ulysse, Ana-Fleur et Alexandre Bonaventure.

Nancy, France

Malte Henkel
Dragi Karevski

Contents

Contributors

Michel Bauer IPhT, CEA-Saclay, Gif-sur-Yvette, France; LPTENS, Paris, France

Christophe Chatelain GPS—DP2M—IJL (CNRS UMR 7198), Université de Lorraine Nancy, Vandœuvre-lès-Nancy Cedex, France

Malte Henkel GPS—DP2M—IJL (CNRS UMR 7198), Université de Lorraine Nancy, Vandœuvre-lès-Nancy Cedex, France

Jesper Lykke Jacobsen Laboratoire de Physique Théorique de l'École Normale Supérieure, Paris, France

Dragi Karevski GPS—DP2M—IJL (CNRS UMR 7198), Université de Lorraine Nancy, Vandœuvre-lès-Nancy Cedex, France

Abbreviations

Acronyms

BA	Bethe ansatz
BCFT	boundary conformal field-theory
CFT	conformal field-theory
CG	Coulomb gas
DLA	diffusion-limited aggregation
FK	Fortuin-Kasteleyn
IRF	interaction-around-a face
ISG	Ising spin glass
LERW	loop-erased random walk
OPA	operator-product algebra
OPE	operator-product expansion
RCFT	rational conformal field-theory
RFIM	random-field Ising model
RG	renormalisation group
SAW	self-avoiding walk
SLE	stochastic Loewner evolution
SOS	solid-on-solid
TL	Temperley-Lieb
$2D$	two-dimensional
$1D$	one-dimensional

Notations[1]

c	Virasoro central charge
κ	SLE parameter or 'diffusion constant'
B_t	standard Brownian motion
g_t	Loewner map
$L_n, \bar{L}_n$	generators of Virasoro algebra
$\Delta, \bar{\Delta}$	conformal weights
$\Delta_{r,s}$	elements of Kac table, conformal weights of minimal models
$x = \Delta + \bar{\Delta}$	scaling dimension
$s = \Delta - \bar{\Delta}$	spin
d_f	fractal dimension
$\mathbf{r}$	spatial coordinates (vector) in $2D$ with components $\mathbf{r} = r_1 + ir_2 = x + iy$
$z, \bar{z}$	complex coordinates
$T_{\mu\nu}$	components of energy-momentum tensor
$T(z), \bar{T}(\bar{z})$	complex energy-momentum tensor
$J(z), \bar{J}(\bar{z})$	conserved chiral current
ϕ, φ	generic symbol for conformal scaling operators
$\chi_{r,s}$	conformal character, of the primary operator $\phi_{r,s}$
V_α	$U(1)$ vertex operator, of charge α
τ	modular parameter of the torus
$P(q)$	generating function for the partitions of the integers related to Dedekind function $\eta(\tau) = q^{1/24} P(q)^{-1}$, with $q = e^{2\pi i \tau}$
Z	partition function, functional integral
$\mathcal{L}$	Lagrangian density of classical field-theory
$\mathcal{H}$	classical hamiltonian (energy) of spin systems
H	quantum hamiltonian, logarithm of transfer matrix
$\langle . \rangle$	thermodynamic average
$\mathbb{Z}, \mathbb{R}, \mathbb{C}; \mathbb{H}$	integer, real & complex numbers; upper complex half-plane
$\mathcal{D}$	complex domain $\subset \mathbb{C}$
$\mathcal{T}$	tiling
Λ	lattice $\subset \mathbb{Z}^d$, with $\mathcal{N} = |\Lambda|$ sites
L	linear size of finite-size domains
$G, \mathfrak{g}$	Lie group and its associated Lie algebra, $\mathfrak{g} = \mathrm{Lie}(G)$

---

[1] Some notations commonly used in this book are listed in this section.

Chapter 1
A Short Introduction to Conformal Invariance

Malte Henkel and Dragi Karevski

1.1 From Scale-Invariance to Conformal Invariance

Conformal invariance arises as an extension of scale-invariance, which is physically realised at a **critical point**, of a many-body system at thermal equilibrium. In order to permit readers to arrive as rapidly as possible at the main themes of this book, we begin with a heuristic guide, leading from critical phenomena to conformal symmetries. For illustrative purposes, the reader might refer to the celebrated **Ising model**, which can be defined in terms of 'spin variables' $\sigma_i = \pm 1$ attached to the sites $\mathbf{i}$ of a hyper-cubic lattice $\Lambda \subset \mathbb{Z}^d$, with $|\Lambda|$ sites. To each configuration of spins $\{\sigma\}$ one associates the energy

$$\mathscr{H}\big[\{\sigma\}\big] = -J \sum_{(\mathbf{i},\mathbf{j})} \sigma_{\mathbf{i}} \sigma_{\mathbf{j}} - h \sum_{\mathbf{i}} \sigma_{\mathbf{i}}. \tag{1.1}$$

Here, the first sum only extends over pairs of nearest neighbours on the lattice Λ, $J > 0$ denotes the exchange coupling and h an external magnetic field. Conventionally, one considers the system being coupled to an external heat bath[1] of temperature T and in principle, one should now try to compute the **partition function** $Z = Z(T,h) := \sum_{\{\sigma\}} e^{-\mathscr{H}[\{\sigma\}]/T}$. If this can be done, the density of the **Gibbs potential** is given by $g(T,h) = -|\Lambda|^{-1} T \ln Z(T,h)$ from which the full thermodynamics can be derived. The study of phase transitions then involves the analysis of possible non-analytic behaviour of g and its derivatives as a function of T or h, in the limit $|\Lambda| \to \infty$.

---

[1] Unless explicitly stated otherwise, we shall choose units such that the Boltzmann constant $k_{\mathrm{B}} = 1$.

M. Henkel (✉) · D. Karevski
GPS—DP2M—IJL (CNRS UMR 7198), Université de Lorraine Nancy, BP 70239,
54506 Vandœuvre-lès-Nancy Cedex, France
e-mail: malte.henkel@ijl.nancy-universite.fr

D. Karevski
e-mail: Dragi.Karevski@ijl.nancy-universite.fr

M. Henkel, D. Karevski (eds.), *Conformal Invariance: an Introduction to Loops, Interfaces and Stochastic Loewner Evolution*, Lecture Notes in Physics 853,
DOI 10.1007/978-3-642-27934-8_1, © Springer-Verlag Berlin Heidelberg 2012

We shall not follow this route, but shall rather illustrate now how phase transitions and critical phenomena can be analysed in terms of correlation functions. Using the Ising model for illustration and going over to a continuum limit, one defines the space-dependent densities $\sigma(\mathbf{r})$ and $\varepsilon(\mathbf{r})$ of the magnetisation and the energy. Then consider the **two-point function**s

$$
\begin{aligned}
G_\sigma(\mathbf{r}_1 - \mathbf{r}_2) &= \langle\sigma(\mathbf{r}_1)\sigma(\mathbf{r}_2)\rangle - \langle\sigma(\mathbf{r}_1)\rangle\langle\sigma(\mathbf{r}_2)\rangle \\
G_\varepsilon(\mathbf{r}_1 - \mathbf{r}_2) &= \langle\varepsilon(\mathbf{r}_1)\varepsilon(\mathbf{r}_2)\rangle - \langle\varepsilon(\mathbf{r}_1)\rangle\langle\varepsilon(\mathbf{r}_2)\rangle
\end{aligned}
\tag{1.2}
$$

where $\langle.\rangle$ denotes the thermodynamic average. From now on, we use the reduced variables $\tau := (T_c - T)/T$ and h and implicitly assume that the model has a critical point at some $T = T_c \neq 0$ and $h = 0$. The hypothesis of **scale-invariance** asserts that under a length rescaling, with a constant rescaling factor b (which generates the **dilatation** $\mathbf{r} \mapsto \mathbf{r}' = \mathbf{r}/b$), the correlators G_σ and G_ε are *generalised homogeneous functions*

$$
\begin{aligned}
G_\sigma(\mathbf{r}; \tau, h) &= b^{-2x_\sigma} G_\sigma\left(\mathbf{r}/b; \tau b^{y_\tau}, h b^{y_h}\right) \\
G_\varepsilon(\mathbf{r}; \tau, h) &= b^{-2x_\varepsilon} G_\varepsilon\left(\mathbf{r}/b; \tau b^{y_\tau}, h b^{y_h}\right).
\end{aligned}
\tag{1.3}
$$

Here x_σ and x_ε are the **scaling dimensions** of σ and ε while y_τ, y_h are usually called **renormalisation-group eigenvalue**s. These numbers are **universal** in the sense that they are *independent* of many details of the models under study (such as the kind of lattice used, or the precise range of the interactions). Their values can be used to characterise the **universality class** of a certain phase transition, which almost always only depends[2] on the space dimension d and the global symmetry of the interaction hamiltonian $\mathcal{H}[\{\sigma\}]$. The densities $\sigma(\mathbf{r})$ and $\varepsilon(\mathbf{r})$ are called **scaling operator**s, while their conjugates h and τ are called **scaling field**s.

The scaling behaviour (1.3) permits to recover the scaling of the thermodynamic observables. Recall that the magnetic susceptibility per spin $\chi = \chi(\tau, h) \simeq -\partial^2 g/\partial h^2$ and the specific heat $C = C(\tau, h) \simeq -\partial^2 g/\partial \tau^2$ are simply related to g, close to criticality. Furthermore, they can also be found from the static **fluctuation-dissipation theorem**s

$$
\chi \simeq \frac{1}{T} \int d^d\mathbf{r}\, G_\sigma(\mathbf{r}), \qquad C \simeq \frac{1}{T^2} \int d^d\mathbf{r}\, G_\varepsilon(\mathbf{r}).
\tag{1.4}
$$

Integrating, one readily finds the **scaling** of the Gibbs potential

$$
g(\tau, h) = b^{-d} g\left(\tau b^{y_\tau}, h b^{y_h}\right),
\tag{1.5}
$$

where we also used $x_\varepsilon + y_\tau = x_\sigma + y_h = d$. This last relation is equivalent to **hyperscaling**. The relationship with the conventional critical exponents α, β, ν can be read off from

$$
x_\sigma = d - y_h = \frac{\beta}{\nu}, \qquad x_\varepsilon = d - y_\tau = \frac{1 - \alpha}{\nu}.
\tag{1.6}
$$

---

[2]For example, in the 2D Ising universality class, one has $x_\sigma = 1/8$ and $x_\varepsilon = 1/2$.

Fig. 1.1 A conformal
transformation (*right panel*)
of a rectangular lattice (*left
panel*)

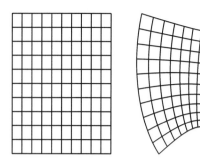

It is left as an easy exercise to re-derive from this the complete critical behaviour
of the specific heat $C(\tau, 0) \sim |\tau|^{-\alpha}$, magnetisation $m(\tau, 0) \sim \tau^{\beta}$, $m(0, h) \sim h^{1/\delta}$,
susceptibility $\chi(\tau, 0) \sim |\tau|^{-\gamma}$, correlation length $\xi \sim -|\mathbf{r}|/\ln G_{\sigma}(\mathbf{r}; \tau, 0) \sim |\tau|^{-\nu}$,
correlator $G(\mathbf{r}; 0, 0) \sim |\mathbf{r}|^{-(d-2+\eta)}$ and so on. Especially, many exponent scaling
relations, such as $\alpha + 2\beta + \gamma = 2$ or $\alpha = 2 - \nu d$, are obtained. Equation (1.5)
implies that only two of these critical exponents are independent.

Scaling operators are cast into three distinct categories: if $x < d$ (or $y = d - x > 0$), the scaling operator is called **relevant**, since under repeated scale-
transformation, the scaling argument $b^{y}\delta \to \infty$. Relevant scaling operators essen-
tially determine the scaling behaviour. If $x > d$ (or $y < 0$), the scaling operator is
called **irrelevant**, since the argument $b^{y}\delta \to 0$ and the corresponding quantity will
not influence the (leading) scaling behaviour. Finally, if $x = d$ (or $y = 0$), the scaling
operator is called **marginal**. The scaling form (1.5) of the Gibbs potential states that
the only two relevant scaling operators are the magnetisation density and the energy
density. This simplifying assumption is indeed realised in many different, but not
all, systems.

In order to see whether more rich scaling symmetries might be possible, we con-
sider generic scaling operators $\phi_a(\mathbf{r})$, arrange for $\langle \phi_a(\mathbf{r}) \rangle = 0$ and study the correlators $G_{ab}(\mathbf{r}_a, \mathbf{r}_b) = \langle \phi_a(\mathbf{r}_a)\phi_b(\mathbf{r}_b) \rangle$. We expect that the transformation properties of
the G_{ab} follow from those of the scaling operators ϕ_a. If we admit space-dependent
rescaling factors $b = b(\mathbf{r})$, a natural generalisation of global scale-invariance (1.3)
is

$$\phi(\mathbf{r}) \mapsto \phi'(\mathbf{r}) = J(\mathbf{r})^{x/d}\phi\big(\mathbf{r}/b(\mathbf{r})\big), \tag{1.7}$$

where $J(\mathbf{r})$ is the Jacobian of the transformation $\mathbf{r} \mapsto \mathbf{r}' = \mathbf{r}/b(\mathbf{r})$ and x the scaling
dimension of ϕ. If one restricts to those coordinate transformations which conserve
angles, one arrives at **conformal transformation**s (see Fig. 1.1).

In this book, we shall only be interested in *two-dimensional (2D) conformal
transformations*. Since the 1980s, starting from the work of Belavin, Polyakov and
Zamolodchikov and of Cardy, the extreme usefulness of conformal invariance in
understanding $2D$ critical phenomena has been realised. In this chapter, we shall try
to compactly introduce the main points and to explain them in a simple way.

In $2D$, it is useful to work with *complex coordinates* (or light-cone coordinates)

$$z = r_1 + ir_2, \qquad \bar{z} = r_1 - ir_2 \tag{1.8}$$

instead of two-dimensional vectors $\mathbf{r} = (r_1, r_2)$. Recall that any analytic (or anti-analytic) coordinate transformation $z \mapsto w(z)$ or $\bar{z} \mapsto \bar{w}(\bar{z})$ is conformal. For technical reasons, we prefer to work with an infinitesimal form and write $w(z) = z + \varepsilon(z)$, with $\varepsilon(z) = -\varepsilon z^{n+1}$. From (1.7), one then obtains the infinitesimal transformations $\delta\phi = \varepsilon\ell_n\phi$ and $\bar{\delta}\phi = \bar{\varepsilon}\bar{\ell}_n\phi$, with $n \in \mathbb{Z}$. Explicitly, they read

$$\ell_n = -z^{n+1}\partial_z - \Delta(n+1)z^n, \qquad \bar{\ell}_n = -\bar{z}^{n+1}\partial_{\bar{z}} - \overline{\Delta}(n+1)\bar{z}^n. \qquad (1.9)$$

Here the non-derivative terms express the transformation of the scaling operator $\phi = \phi(z, \bar{z})$ and the terms containing derivatives describe the changes in the coordinates z and $\bar{z}$. The real numbers Δ and $\overline{\Delta}$ are called **conformal weight**s of the scaling operator ϕ. They are related to the scaling dimension x and spin s as follows[3]

$$x = \Delta + \overline{\Delta}, \qquad s = \Delta - \overline{\Delta}. \qquad (1.10)$$

The generators (1.9) satisfy the following commutation relations

$$[\ell_n, \ell_m] = (n - m)\ell_{m+n}$$
$$[\bar{\ell}_n, \bar{\ell}_m] = (n - m)\bar{\ell}_{m+n} \qquad (1.11)$$
$$[\ell_n, \bar{\ell}_m] = 0.$$

This algebra is also called the **loop algebra** (or centre-less Virasoro algebra)[4] and decomposes into the direct sum of two commuting Lie algebras, one generated by the set $\langle\ell_n\rangle_{n\in\mathbb{Z}}$ and the other by $\langle\bar{\ell}_n\rangle_{n\in\mathbb{Z}}$. Because of this simple structure, it is often enough to consider merely the z-dependence of correlators.

Exercise 1 Verify that the two-dimensional Laplace operator $\nabla^2 = \partial_{r_1}^2 + \partial_{r_2}^2 = 4\frac{\partial}{\partial z}\frac{\partial}{\partial\bar{z}}$ satisfies

$$\left[\nabla^2, \ell_n\right] = -(n+1)z^n\nabla^2 - 4\Delta(n+1)nz^{n-1}\frac{\partial}{\partial\bar{z}}.$$

Show that the 2D Laplace equation $\nabla^2\phi(z, \bar{z}) = 0$ is conformally invariant, if the conformal weights $\Delta = \overline{\Delta} = 0$.

The Lie algebra (1.11) is infinite-dimensional. Its maximal finite-dimensional sub-algebra consists of the **projective conformal transformations** and is given by $\mathfrak{sl}(2, \mathbb{R}) \oplus \mathfrak{sl}(2, \mathbb{R}) = \langle\ell_n\rangle_{n=\pm1,0} \oplus \langle\bar{\ell}_n\rangle_{n=\pm1,0}$. This subset of conformal transformations is the only one which has analogues in $d > 2$ dimensions and which

---

[3]In the literature, the alternative notation: $h, \bar{h}$ for the conformal weights and $\Delta = h + \bar{h}$ for the scaling dimension, is also met with frequently.

[4]Although this algebra was first defined by É. Cartan in 1909, it is unfortunately often referred to as 'Witt algebra'. Witt studied this algebra only much later (in the 1930s), over fields of characteristic $p > 0$, when the algebra is spanned by the ℓ_n with $-1 \leq n \leq p - 2$.

map the entire complex plane $\mathbb{C}$ (including the point at infinity) onto itself. Explicitly, these are translations, generated by $\ell_1, \bar{\ell}_{-1}$, dilatations $\ell_0 + \bar{\ell}_0$, rotations $i(\ell_0 - \bar{\ell}_0)$ and special conformal transformations $\ell_1, \bar{\ell}_1$. One calls **quasi-primary** those scaling operators ϕ which transform co-variantly, that is according to (1.9), under $\mathfrak{sl}(2, \mathbb{R}) \oplus \mathfrak{sl}(2, \mathbb{R})$. Correlators built exclusively from quasi-primary scaling operators $\phi_a = \phi_a(z_a, \bar{z}_a)$ satisfy the **projective Ward identities**

$$\sum_{i=1}^{n} \frac{\partial}{\partial z_i} \langle \phi_1 \ldots \phi_n \rangle = 0, \qquad \sum_{i=1}^{n} \left(z_i \frac{\partial}{\partial z_i} + \Delta_i \right) \langle \phi_1 \ldots \phi_n \rangle = 0,$$

$$\sum_{i=1}^{n} \left(z_i^2 \frac{\partial}{\partial z_i} + 2 \Delta_i z_i \right) \langle \phi_1 \ldots \phi_n \rangle = 0. \tag{1.12}$$

A similar set of equations holds for the dependence on the variables $\bar{z}_i$. The projective Ward identities express the vanishing of the n-particle extensions of the generators $\ell_{\pm 1, 0}$ from (1.9) on the co-variant n-point correlators $\langle \phi_1 \ldots \phi_n \rangle$.

We illustrate how the conditions (1.12) determine the form of the conformally covariant two-point function $\Phi(z_1, z_2; \bar{z}_1, \bar{z}_2) := \langle \phi_1(z_1, \bar{z}_1) \phi_2(z_2, \bar{z}_2) \rangle$. It is enough to study the dependence on z_1 and z_2 explicitly. From translation invariance, it is clear that $\Phi = \Phi(z)$ with $z = z_1 - z_2$. Next, scale-invariance implies

$$\ell_0 \Phi(z) = (-z \partial_z - \Delta_1 - \Delta_2) \Phi(z) = 0 \tag{1.13}$$

with the solution $\Phi(z) = \Phi_0 z^{-\Delta_1 - \Delta_2}$. Finally, invariance under the special transformation gives

$$\ell_1 \Phi(z) = \left(-(z_1^2 - z_2^2) \partial_z - 2\Delta_1 z_1 - 2\Delta_2 z_2 \right) \Phi(z)$$
$$= \left(-z^2 \partial_z - 2\Delta_1 z \right) \Phi(z) + 2z_2 \underbrace{(-z \partial_z - \Delta_1 - \Delta_2) \Phi(z)}_{=0} = 0, \tag{1.14}$$

where we used the decomposition $z_1^2 - z_2^2 = (z_1 - z_2)^2 + 2z_2(z_1 - z_2)$. The last term in the second line of (1.14) vanishes because of dilatation-invariance (1.13). Next, multiply (1.13) by z and subtract it from (1.14). This leads to

$$(\Delta_1 - \Delta_2) z \Phi(z) = 0. \tag{1.15}$$

Therefore, the conformal weights of the two scaling operators have to be equal. Combining these results and restoring the conjugate part as well, the two-point function of quasi-primary scaling operators $\phi_{a,b}$ must be

$$\langle \phi_a(z_a, \bar{z}_a) \phi_b(z_b, \bar{z}_b) \rangle = \delta_{\Delta_a, \Delta_b} \delta_{\bar{\Delta}_a, \bar{\Delta}_b} \phi_{ab} (z_a - z_b)^{-2\Delta_a} (\bar{z}_a - \bar{z}_b)^{-2\bar{\Delta}_a}, \tag{1.16}$$

where ϕ_{ab} is an arbitrary normalisation constant. The constraint on the conformal weights goes beyond what is found from scale-invariance alone.

Exercise 2 Derive the conformally invariant **three-point function** of quasi-primary scaling operators[5] (with $z_{ab} := z_a - z_b$)

---

[5] With the normalisation $\phi_{ab} = \delta_{ab}$ in (1.16), the coefficient $\mathscr{C}_{123}$ is universal and *not* arbitrary.

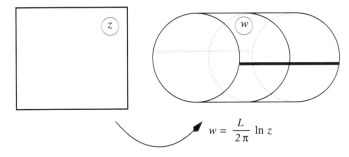

$$w = \frac{L}{2\pi} \ln z$$

Fig. 1.2 Conformal transformation $w = w(z) = \frac{L}{2\pi} \ln z$ from the infinite complex plane $\mathbb{C}$ to the infinitely long strip of finite width L and periodic boundary conditions, indicated by the seam

$$\langle \phi_1 \phi_2 \phi_3 \rangle = \mathscr{C}_{123} z_{12}^{-(\Delta_1+\Delta_2-\Delta_3)} z_{23}^{-(\Delta_2+\Delta_3-\Delta_1)} z_{13}^{-(\Delta_1+\Delta_3-\Delta_2)}$$

$$\times \bar{z}_{12}^{-(\overline{\Delta}_1+\overline{\Delta}_2-\overline{\Delta}_3)} \bar{z}_{23}^{-(\overline{\Delta}_2+\overline{\Delta}_3-\overline{\Delta}_1)} \bar{z}_{13}^{-(\overline{\Delta}_1+\overline{\Delta}_3-\overline{\Delta}_2)}. \quad (1.17)$$

The form of higher correlators cannot be fully determined from the projective Ward identities.

It is a special feature of $2D$ conformal invariance that it can relate correlators in different geometries. This goes beyond the finite-dimensional sub-algebra considered so far. It becomes necessary to sharpen the terminology introduced so far, by defining a **primary** scaling operator ϕ, with conformal weights $\Delta, \overline{\Delta}$, by the transformation law, under the conformal transformation $z \mapsto w = w(z)$ and $\bar{z} \mapsto \bar{w} = \bar{w}(\bar{z})$

$$\phi(z, \bar{z}) = \left(\frac{dw(z)}{dz}\right)^{\Delta} \left(\frac{d\bar{w}(\bar{z})}{d\bar{z}}\right)^{\overline{\Delta}} \phi(w, \bar{w}). \quad (1.18)$$

The most commonly met example is given by the logarithmic conformal transformation

$$w = \frac{L}{2\pi} \ln z \quad (1.19)$$

which maps the infinite complex z-plane onto an infinitely long strip of finite width L, with *periodic* boundary conditions. In Fig. 1.2, this strip becomes the surface of the infinitely long cylinder of circumference L and the thick line indicates the seam. A two-point function built from a primary scaling operator ϕ then transforms as

$$\langle \phi(z_1, \bar{z}_1) \phi(z_2, \bar{z}_2) \rangle_z$$

$$= \left(\frac{dw}{dz}(z_1) \frac{dw}{dz}(z_2)\right)^{\Delta} \left(\frac{d\bar{w}}{d\bar{z}}(\bar{z}_1) \frac{d\bar{w}}{d\bar{z}}(\bar{z}_2)\right)^{\overline{\Delta}} \langle \phi(w_1, \bar{w}_1) \phi(w_2, \bar{w}_2) \rangle_w. \quad (1.20)$$

Using the explicit form (1.16) of the two-point function in the z-plane, one obtains for the transformation (1.19), with $z = \exp(2\pi L^{-1} w) = \exp(2\pi L^{-1}(u + iv))$

$$\left\langle\phi(w_1,\bar{w}_1)\phi(w_2,\bar{w}_2)\right\rangle_w$$

$$= \left(\frac{2\pi}{L}\right)^{2\Delta+2\bar{\Delta}}\left(\frac{z_1^{1/2}z_2^{1/2}}{z_1-z_2}\right)^{2\Delta}\left(\frac{\bar{z}_1^{1/2}\bar{z}_2^{1/2}}{\bar{z}_1-\bar{z}_2}\right)^{2\bar{\Delta}}$$

$$= \left(\frac{2\pi}{L}\frac{\exp[\frac{\pi}{L}(w_1+w_2)]}{\exp(\frac{2\pi}{L}w_1)-\exp(\frac{2\pi}{L}w_2)}\right)^{2\Delta}\cdot\left(\frac{2\pi}{L}\frac{\exp[\frac{\pi}{L}(\bar{w}_1+\bar{w}_2)]}{\exp(\frac{2\pi}{L}\bar{w}_1)-\exp(\frac{2\pi}{L}\bar{w}_2)}\right)^{2\bar{\Delta}}$$

$$= \left(\frac{\pi}{L}\frac{1}{\sinh[\frac{\pi}{L}(w_1-w_2)]}\right)^{2\Delta}\cdot\left(\frac{\pi}{L}\frac{1}{\sinh[\frac{\pi}{L}(\bar{w}_1-\bar{w}_2)]}\right)^{2\bar{\Delta}}, \tag{1.21}$$

where $w_1-w_2=(u_1-u_2)+i(v_1-v_2)$. Evidently, in the limit $|w_1-w_2|\ll L$, one simply recovers the two-point function (1.16) in the complex plane $\mathbb{C}$. On the other hand, and taking $v_1=v_2$, the opposite limit $|u_1-u_2|\gg L$ gives the asymptotic exponential decay

$$\left\langle\phi(u_1,0),\phi(u_2,0)\right\rangle_{\text{strip}}\simeq\left(\frac{2\pi}{L}\right)^{2x}\exp\left[-\frac{2\pi}{L}(\Delta+\bar{\Delta})(u_1-u_2)\right], \tag{1.22}$$

which is the usual way to define a correlation length ξ, via $\langle\phi(u,0)\phi(0,0)\rangle\sim\exp(-u/\xi)$. Because of $x=\Delta+\bar{\Delta}$, one reads off

$$\xi=L/(2\pi x). \tag{1.23}$$

Therefore, conformal invariance has produced a prediction for the **finite-size scaling** of the correlation length in a *finite* geometry. Remarkably, it relates the finite-size scaling amplitude to an *universal* scaling dimension. The celebrated result Eq. (1.23) is an often-used practical way to measure scaling dimensions in specific models.

Exercise 3 If ϕ is a quasi-primary scaling operator, how does its derivative $\partial_z\phi$ transform?

1.2 The Energy-Momentum Tensor

More information on conformal field-theories comes from an analysis of the transformations of averages. Following well-known procedures, such as the Hubbard-Stratanovich transformation, one may associate to a classical spin system, such as the Ising model (defined on a discrete lattice Λ) and described by a classical hamiltonian $\mathcal{H}[\{\sigma\}]$, a continuum field-theory, which in turn is characterised by an action $S[\phi]$. Then, for an observable $\mathcal{A}$

$$\langle\mathcal{A}\rangle=\frac{1}{Z}\sum_{\{\sigma\}}\mathcal{A}[\{\sigma\}]e^{-\mathcal{H}[\{\sigma\}]/T}=\frac{1}{Z}\int\mathcal{D}\phi\,\mathcal{A}[\phi]\,e^{-S[\phi]}, \tag{1.24}$$

where $Z=\sum_{\{\sigma\}}e^{-\mathcal{H}[\{\sigma\}]/T}=\int\mathcal{D}\phi\,e^{-S[\phi]}$ is the partition function and $\mathcal{D}\phi=\prod_{\mathbf{r}}\int d\phi(\mathbf{r})$ serves here a shorthand for the functional integration over the values of the continuum field at all space points and which formally replaces the sum over all spin configurations in the lattice model.

The specific form of the action is in general quite complicated. However, it is usually admitted that the Ising universality class can be described in terms of a scalar ϕ^4-theory, with $S = \int d^d r \, \mathscr{L}$ and $\mathscr{L} = \frac{1}{2}(\nabla\phi)^2 + \frac{m^2}{2}\phi^2 + \frac{g}{4}\phi^4$. At the level of mean-field theory, g is a constant and $m^2 \sim T - T_c$.

Now, consider how this transforms under a coordinate transformation $\mathbf{r} \mapsto \mathbf{r}' := \mathbf{r} + \varepsilon(\mathbf{r})$. A critical system, where the control parameters $\tau = h = 0$, should be at a renormalisation group fixed point and be described by a fixed point action $S^*[\phi]$. A generic coordinate transformation will in general not be conformal, hence it will contain shear components which will drive the system away from its critical point, such that the action transforms $S^*[\phi] \to S^*[\phi] + \delta S[\phi]$. For an infinitesimal transformation, one expects to first order, since the action is a scalar under rotations

$$\delta S = -\frac{1}{I_d} \int d^d \mathbf{r} \, \partial^\mu \varepsilon^\nu(\mathbf{r}) T_{\mu\nu}(\mathbf{r}), \tag{1.25}$$

where $\mu, \nu = 1, \ldots, d$, Einstein's summation convention is used, I_d is a constant ($I_2 = 2\pi$) and $T_{\mu\nu}$ is the **energy-momentum tensor**. Besides the obviously assumed spatial translation-invariance, we shall not enter into a detailed discussion on the validity of Eq. (1.25), and shall rather regard it as a *postulate* which selects the kind of 'local' field-theories we wish to study.[6] This implies the **conformal Ward identity** for the n-point functions of quasi-primary scaling operators

$$\sum_{p=1}^{n} \left\langle \phi_1(\mathbf{r}_1) \ldots \left(\varepsilon(\mathbf{r}_p) \cdot \nabla + \frac{x_p}{d} \nabla \cdot \varepsilon(\mathbf{r}_p) \right) \phi_p(\mathbf{r}_p) \ldots \phi_n(\mathbf{r}_n) \right\rangle$$

$$+ \frac{1}{I_d} \int d\mathbf{r} \left\langle \phi_1(\mathbf{r}_1) \ldots \phi_n(\mathbf{r}_n) T_{\mu\nu}(\mathbf{r}) \right\rangle \partial^\mu \varepsilon^\nu(\mathbf{r}) = 0. \tag{1.26}$$

This is the fundamental equation for conformal field-theories. For projective conformal transformations, the second line is absent and one is back to (1.12).

Proof In view of the importance of (1.26), we outline a proof. Recalling (1.7), averages transform as

$$\left\langle \phi_1'(\mathbf{r}_1) \ldots \phi_n'(\mathbf{r}_n) \right\rangle_{S^*} = J(\mathbf{r}_1)^{x_1/d} \ldots J(\mathbf{r}_n)^{x_n/d} \left\langle \phi_1(\mathbf{r}_1') \ldots \phi_n'(\mathbf{r}_n') \right\rangle_{S^*+\delta S}.$$

To leading order, one has

$$\left\langle \mathscr{A} \right\rangle_{S^*+\delta S} \simeq \frac{1}{Z} \int \mathscr{D}\phi \mathscr{A}[\phi] \left(1 - \delta S[\phi] \right) e^{-S^*[\phi]} = \left\langle \mathscr{A}(1 - \delta S) \right\rangle_{S^*}, \tag{1.27}$$

whereas the coordinate change simply amounts to a change of variables in Z. From the Jacobian, one has the explicit transformation

$$\delta\phi(\mathbf{r}) = \left(\frac{x}{d} \nabla \cdot \varepsilon(\mathbf{r}) + \varepsilon(\mathbf{r}) \cdot \nabla \right) \phi(\mathbf{r}) \tag{1.28}$$

---

[6]Equation (1.25) borrows from the theory of elasticity, where $T_{\mu\nu}$ is called 'stress-energy tensor'. This analogy is unlikely to be valid for theories with long-range interactions.

of a quasi-primary scaling operator ϕ. If one requires conformal invariance, one expects

$$\sum_{p=1}^{n}\langle\phi_1(\mathbf{r}_1)\ldots\delta_{\text{total}}\phi_p(\mathbf{r}_p)\ldots\phi_n(\mathbf{r}_n)\rangle = 0$$

where $\delta_{\text{total}}\phi$ is the total change in ϕ, either explicitly via (1.28) or through a change of the measure via (1.27). Combination gives the announced result. □

Several important properties of $T_{\mu\nu}$ follow from the conformal Ward identity.

1. *Conservation law* $\partial^\mu T_{\mu\nu} = 0$, which follows from (1.26) by partial integration.
2. Spatial translation-invariance has been admitted from the beginning.
3. Invariance under *rotations*, with $\varepsilon^\mu(\mathbf{r}) = \varepsilon^{\mu\nu}r_\nu$ and $\varepsilon^{\mu\nu} = -\varepsilon^{\nu\mu}$, implies a symmetric tensor, $T_{\mu\nu} = T_{\nu\mu}$.
4. *Dilatation*-invariance, with $\varepsilon^\mu(\mathbf{r}) = \lambda r^\mu$ and λ a constant, implies tracelessness, $T_\mu^\mu = 0$.

These conclusions are obtained through partial integrations, and discarding any boundary terms. If that procedure is legitimate, such that $T_{\mu\nu}$ is symmetric and traceless, conformal invariance would follows. Schematically,

$$\left.\begin{array}{l}\text{translation-invariance}\\\text{rotation-invariance}\\\text{scale (dilatation)-invariance}\\\text{conformal Ward identity}\end{array}\right\} \Longrightarrow \text{conformal invariance.} \qquad (1.29)$$

To see this formally, consider an infinitesimal special conformal transformation, with $\varepsilon^\mu(\mathbf{r}) = \eta^\mu \mathbf{r}^2 - 2r^\mu\eta\cdot\mathbf{r}$ and where η is some constant infinitesimal vector. Hence it follows that $T_{\mu\nu}\partial^\mu\varepsilon^\nu = 2T_{\mu\nu}(r^\mu\eta^\nu - \eta^\mu r^\nu) - 2T_\mu^\mu\eta\cdot\mathbf{r} = 0$, which implies $\delta S = 0$.

We refer to the literature [2, 28, 37, 38] for careful discussions, including counter-examples (!), on the subtle hidden assumptions required for the validity of these conclusions. We point out that in $2D$, this formal argument can be extended to the full infinite-dimensional conformal algebra. If one of the conditions in (1.29) is not met, full conformal invariance will not hold.

Exercise 4 In classical field-theories described by a lagrangian density $\mathscr{L}$, the classical equations of motion are $\partial_\mu\partial\mathscr{L}/\partial(\partial_\mu\phi) - \partial\mathscr{L}/\partial\phi = 0$. In euclidean d-dimensional space, one usually considers the '*canonical energy-momentum tensor*'

$$\widetilde{T}_{\mu\nu} := \frac{\partial\mathscr{L}}{\partial(\partial_\mu\phi)}\partial_\nu\phi - \delta_{\mu\nu}\mathscr{L}.$$

Consider a free field with the lagrangian $\mathscr{L} = \frac{1}{2}(\partial_\mu\phi)^2$. Using the classical equations of motion, show that it is *not* $\widetilde{T}_{\mu\nu}$, but rather the *improved* energy-momentum tensor

$$T_{\mu\nu} = \partial_\mu\phi\partial_\nu\phi - \frac{1}{2}\delta_{\mu\nu}(\partial_\lambda\phi)^2 + \frac{1}{4}\frac{d-2}{d-1}\left(\delta_{\mu\nu}\nabla^2 - \partial_\mu\partial_\nu\right)\phi^2 \qquad (1.30)$$

Fig. 1.3 *Contour C which encloses the n points $z_1, \ldots, z_n$, for the derivation of the $2D$ conformal Ward identity (1.32). The interior domain $D_1 = \mathrm{int}\, C$ is shaded*

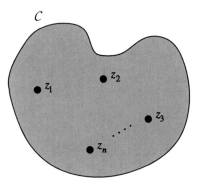

which satisfies all required properties of an energy-momentum tensor of a conformal field-theory.

Considerably more can be said in $2D$. Since $T_{\mu\nu}$ is symmetric and traceless, it has two independent components. In complex coordinates these may be taken to be

$$T := T_{zz} = \frac{1}{2}(T_{11} - \mathrm{i}T_{12}), \qquad \bar{T} := \bar{T}_{\bar{z}\bar{z}} = \frac{1}{2}(T_{11} + \mathrm{i}T_{12}). \qquad (1.31)$$

The conservation law then gives $\partial_{\bar{z}} T = \partial_z \bar{T} = 0$, which are the Cauchy-Riemann equations for $T = T(z)$ being complex analytic in z and $\bar{T} = \bar{T}(\bar{z})$ being anti-analytic.

Returning to the conformal Ward identity, one must take **Liouville's theorem** into account, which states that *a complex function $w : \mathbb{C} \to \mathbb{C}$, analytic in the entire complex plane $\mathbb{C}$ and which is bounded in $\mathbb{C}$, that is there is a constant $M < \infty$ such that $|w(z)| < M$, must be a constant.* Hence, an analytic function $\varepsilon(z)$ with $|\varepsilon(z)| \ll 1$ everywhere in $\mathbb{C}$ cannot exist. In order to rewrite the conformal Ward identity (1.26) in complex coordinates, consider a contour C which encloses the points $z_1, \ldots, z_n$, see Fig. 1.3. In the interior domain $D_1 := \mathrm{int}\, C$, the function $\varepsilon(z)$ is both analytic and small, whereas in the exterior domain $D_2 = \mathbb{C} - D_1$, $\varepsilon(z)$ is small, but not everywhere analytic. If the scaling operators $\phi_1, \ldots, \phi_n$ are all primary, it can then be shown by the residue theorem that the Ward identity (1.26) takes the form

$$\frac{1}{2\pi \mathrm{i}} \oint_C \mathrm{d}z \, \varepsilon(z) \langle T(z)\phi_1(z_1, \bar{z}_1) \ldots \phi_n(z_n, \bar{z}_n) \rangle$$
$$= \frac{1}{2\pi \mathrm{i}} \oint_C \mathrm{d}z \, \varepsilon(z) \sum_{p=1}^{n} \left(\frac{\Delta_p}{(z - z_p)^2} + \frac{1}{z - z_p}\frac{\partial}{\partial z_p} \right) \langle \phi_1(z_1, \bar{z}_1) \ldots \phi_n(z_n, \bar{z}_n) \rangle. \qquad (1.32)$$

The second line may also be interpreted as the change $\delta_\varepsilon \langle \phi_1 \ldots \phi_n \rangle$ of the n-point function under a conformal transformation.

Proof Equation (1.32) can be proven as follows. Denote the first term in (1.26) by $\delta_\varepsilon \langle \phi_1 \ldots \phi_n \rangle$. Then (c.c. denotes the complex conjugate)

$$\delta_\varepsilon \langle \phi_1 \ldots \phi_n \rangle$$

$$= -\int_{D_2} \frac{d\bar{z} \wedge dz}{2\pi i} \partial_{\bar{z}} \varepsilon(z, \bar{z}) \langle T(z) \phi_1(z_1, \bar{z}_1) \ldots \phi_n(z_n, \bar{z}_n) \rangle + \text{c.c.}$$

$$= \int_{D_2} \frac{d\bar{z} \wedge dz}{2\pi i} \varepsilon(z, \bar{z}) \partial_{\bar{z}} \langle T(z) \phi_1(z_1, \bar{z}_1) \ldots \phi_n(z_n, \bar{z}_n) \rangle + \text{c.c.}$$

$$= \int_{D_1} \frac{d\bar{z} \wedge dz}{2\pi i} \varepsilon(z) \partial_{\bar{z}} \langle T(z) \phi_1(z_1, \bar{z}_1) \ldots \phi_n(z_n, \bar{z}_n) \rangle + \text{c.c.}$$

$$= \oint_{\partial D_1} \frac{dz}{2\pi i} \varepsilon(z) \langle T(z) \phi_1(z_1, \bar{z}_1) \ldots \phi_n(z_n, \bar{z}_n) \rangle + \text{c.c.}$$

$$= \sum_{j=1}^{n} \oint_{C_j} \frac{dz}{2\pi i} \varepsilon(z) \langle T(z) \phi_1(z_1, \bar{z}_1) \ldots \phi_n(z_n, \bar{z}_n) \rangle + \text{c.c.} \qquad (1.33)$$

where in the domain D_2, $\varepsilon(z, \bar{z})$ is not necessarily analytic, but infinitesimal small so that the boundary term coming from the partial integration is negligible. Since all the singularities of the integrand only occur at $z = z_j$, the integration domain may be shrunk to D_1, where $\varepsilon = \varepsilon(z)$ is analytic, i.e. $\partial_{\bar{z}} \varepsilon = 0$. Then the complex integration over $\bar{z}$ can be reduced to the boundary, such that only the contour integral over the boundary $C = \partial D_1$ remains. In the last step, this integral is separated into distinct contour integrals over small circles C_j around each point z_j. On the other hand, if ϕ is primary, it transforms as

$$\delta_\varepsilon \phi(z_1, \bar{z}_1) = \Delta \varepsilon'(z_1) \phi(z_1, \bar{z}_1) + \varepsilon(z_1) \partial \phi(z_1)$$

$$= \frac{1}{2\pi i} \oint_{C_1} dz \, \varepsilon(z) \left(\frac{\Delta}{(z - z_1)^2} \phi(z_1, \bar{z}_1) + \frac{1}{z - z_1} \partial_{z_1} \phi(z_1, \bar{z}_1) \right). \quad (1.34)$$

One sums over the contributions of all the $\phi_1, \ldots, \phi_n$ and turns the sum of the contour integrals $\oint_{C_j}$ into a single contour integral $\oint_C$. Joining this with the relation (1.33) gives the assertion. □

Since Eq. (1.32) holds true for an infinitely large ensemble of small functions $\varepsilon(z)$, one can go over from this integral form to a *local* form of the **conformal Ward identity**

$$\langle T(z) \phi_1(z_1, \bar{z}_1) \ldots \phi_n(z_n, \bar{z}_n) \rangle$$

$$= \sum_{p=1}^{n} \left(\frac{\Delta_p}{(z - z_p)^2} + \frac{1}{z - z_p} \frac{\partial}{\partial z_p} \right) \langle \phi_1(z_1, \bar{z}_1) \ldots \phi_n(z_n, \bar{z}_n) \rangle. \quad (1.35)$$

Clearly, an analogous form holds true for $\bar{T}(\bar{z})$, and is obtained from (1.35) by replacing $z \mapsto \bar{z}$, $z_p \mapsto \bar{z}_p$ and $\Delta_p \mapsto \overline{\Delta}_p$. It is important here that the scaling operators are all primary, and as we shall see shortly, Eq. (1.35) does not hold if this condition is not met. The main feature is the *singular* behaviour when $z \to z_p$. The two terms therein have a clear meaning. The first one describes the change in the primary operator ϕ and comes from the Jacobian in the transformation law (1.7). The second one describes the change in the coordinates and reflects the fact that the energy-momentum tensor T is the infinitesimal generator of conformal transformations. These two terms give the most simple way a scaling operator can transform.

One isolates this singular behaviour in defining an **operator product expansion** (OPE) of the energy-momentum tensor T with a primary operator ϕ

$$T(z)\phi(z_1,\bar{z}_1) = \left(\frac{\Delta}{(z-z_1)^2} + \frac{1}{z-z_1}\frac{\partial}{\partial z_1} \right)\phi(z_1,\bar{z}_1) + \text{regular terms} \qquad (1.36)$$

where the so-called 'regular terms', whose average vanishes, are non-singular when $z \to z_1$.

How does the energy-momentum tensor $T(z)$ transform under a conformal transformation? One may characterise it by the averages

$$\langle T(z)\rangle = 0, \qquad \langle T(z_1)T(z_2)\rangle = \frac{c/2}{(z_1-z_2)^4} \qquad (1.37)$$

where one has fixed the energy scale and one has the conformal weights $\Delta_T = 2$ and $\overline{\Delta}_T = 0$, as already suggested from the free-field expression (1.30). Since the energy-momentum tensor is conserved, the well-known Okubo-Fubini-Furlan-Takahashi **non-renormalisation theorem**s [7] from quantum field-theory assert that its conformal weights retain their values from classical, non-renormalised field-theory. The universal constant c, which will turn out to be the single most important constant in conformal field-theory, is called the **central charge**. T generates the following change onto itself, where C_z is a simple contour around the point z

$$\delta_\varepsilon\langle T(z)\rangle = \frac{1}{2\pi i}\oint_{C_z} dz'\,\varepsilon(z')\langle T(z')T(z)\rangle$$

$$= \frac{1}{2\pi i}\oint_{C_z} dz'\,\varepsilon(z')\frac{c/2}{(z'-z)^4}$$

$$= \frac{c}{12}\varepsilon'''(z) \qquad (1.38)$$

by the residue theorem. Since for $c \neq 0$ this is non-vanishing for a non-projective transformation, $T(z)$ is a quasi-primary, but *not* a primary scaling operator. For a non-vanishing central charge, the infinite-dimensional Lie algebra of analytic conformal transformation is broken down to the projective conformal transformations $\mathfrak{sl}(2,\mathbb{R})$. Such a fluctuation-induced symmetry-breaking is called an **anomaly**, and c is sometimes referred to as 'conformal anomaly number'. The equivalent OPE reads

$$T(z)T(z')$$
$$= \frac{c/2}{(z-z')^4} + \frac{2}{(z-z')^2}T(z') + \frac{1}{z-z'}\frac{\partial}{\partial z'}T(z') + \text{regular terms} \qquad (1.39)$$

and one recognises, besides the two expected terms for any scaling operator, a further singular contribution. This extra term distinguishes the OPE of T from the OPE (1.36) of a primary operator. It is possible to integrate the above infinitesimal transformation of $T(z)$. The calculation being tedious and un-inspiring, we merely quote the transformation law under a conformal transformation $w = w(z)$

$$T(z) = T(w) \left(\frac{dw}{dz} \right)^2 + \frac{c}{12} \{w, z\} \tag{1.40}$$

and where one also needs the **Schwarzian derivative**

$$\{w, z\} := \frac{w'''(z)}{w'(z)} - \frac{3}{2} \left(\frac{w''(z)}{w'(z)} \right)^2. \tag{1.41}$$

Exercise 5 For a projective conformal transformation $w(z) = \frac{\alpha z + \beta}{\gamma z + \delta} \in \mathfrak{sl}(2, \mathbb{R})$ with $\alpha\delta - \beta\gamma = 1$, verify that $\{w, z\} = 0$.

Exercise 6 Use the logarithmic transformation (1.19) to show that on an infinitely long strip of finite width L and with periodic boundary conditions, the energy-momentum tensor becomes

$$T_{\text{strip}}(w) = \left(\frac{2\pi}{L} \right)^2 \left(T_{\text{plane}}(z) z^2 - \frac{c}{24} \right). \tag{1.42}$$

In particular, (1.42) implies $\langle T_{\text{strip}}(w) \rangle = -\frac{c}{24} \left(\frac{2\pi}{L} \right)^2$. This is the analogue of the **Casimir effect**, which originally described the lowering of the vacuum energy between two large parallel conducting plates at a finite distance. To make this more explicit in our context, write $w = u + iv$ for the complex coordinate on the strip. The energy operator (or **quantum hamiltonian**) is obtained by integrating over a section of the strip

$$H = \frac{1}{2\pi} \int_0^L dv \, T_{uu}(v) = \frac{1}{2\pi} \int_0^L dv \, (T_{\text{strip}}(w) + \bar{T}_{\text{strip}}(\bar{w})) \tag{1.43}$$

such that the ground-state (vacuum) energy of the hamiltonian H becomes

$$E_0 = \frac{1}{2\pi} \underbrace{\left(\frac{2\pi}{L} \right)^2 \left(-\frac{c}{24} \right)}_{\langle T_{\text{strip}}(w) \rangle} \times 2 \times L = -\frac{\pi c}{6} \frac{1}{L}. \tag{1.44}$$

In our statistical mechanics context, the partition function can be expressed through a trace over the transfer matrix $\mathscr{T}$, according to $Z = \text{tr} \, \mathscr{T}^L$. Since $\mathscr{T}$ plays the rôle of a time-evolution operator at imaginary times ϑ, one can introduces a quantum hamiltonian via $\mathscr{T} = e^{-\vartheta H}$, such that $Z = \text{tr} \, e^{-L\vartheta H}$. In this way, one has a correspondence between a classical $2D$ statistical mechanics model, described by the classical hamiltonian $\mathscr{H}$, and an one-dimensional quantum system described by the quantum hamiltonian H. We shall draw two conclusions from this correspondence:

1. For a $2D$ classical system, at its critical point, on the strip of finite width L, (1.44) gives the following expression for the density of the Gibbs potential

$$g = g_{\text{bulk}} - \frac{\pi c}{6} L^{-2} + o(L^{-2}) \tag{1.45}$$

where the non-universal bulk contribution g_{bulk} was eliminated from our discussion by the first condition (1.37). Equation (1.45) describes a universal finite-size effect for the critical Gibbs potential and furnishes a very efficient and often-used algorithm for the determination of the value of the central charge c in a given model. For example, in the $2D$ Ising universality class one has $c = \frac{1}{2}$.

2. One may equally well consider an one-dimensional quantum system defined by the quantum hamiltonian H. From the above correspondence, the width L of the strip can be identified with the temperature T of the quantum system, via $L^{-1} \leftrightarrow k_B T$. Similarly, the ground state energy is related to the Gibbs potential of the quantum system via $E_0(L) \leftrightarrow g/(k_B T)$. This correspondence depends for its validity on the necessary condition of a linear energy-momentum dispersion relation $E(p) \simeq v_s |p|$, at least in the low-momentum range $|p| \to 0$, and where v_s is the speed of sound in the quantum system.

   If we choose units such that $v_s = 1$, one reads off from (1.44) the Gibbs potential density $g = -(\pi c/6)(k_B T)^2 + o(T^2)$. Hence, the specific heat of the $1D$ quantum system becomes, for $T \to 0$

$$C = -T \frac{\partial^2 g}{\partial T^2} = \frac{\pi c}{3} k_B^2 T \qquad (1.46)$$

which allows to measure the value of c in $1D$ quantum systems with a linear dispersion relation.[7]

Example Consider the ideal $1D$ Bose gas. With the speed of sound $v_s = 1$, one has the linear dispersion relation $\varepsilon_p = |p|$ and the number of particles at momentum p is given by $n_p = (\exp(\varepsilon_p/k_B T) - 1)^{-1}$. Hence the total energy becomes

$$E = \sum_p \varepsilon_p n_p \simeq \frac{1}{2\pi} \int_0^\infty dp\, |p| \big(e^{|p|/(k_B T)} - 1\big)^{-1} \times 2 = \frac{\pi}{6} k_B^2 T^2$$

and finally the specific heat $C = \partial E/\partial T = \frac{\pi}{3} k_B^2 T$. Comparison with Eq. (1.46) gives the central charge $c = 1$ for the $1D$ free boson.

Exercise 7 Show that $c = \frac{1}{2}$ for the $1D$ ideal Fermi gas.

Finally, we remark that c may be viewed as a measure of the importance of fluctuation effects at the critical point. For example, mean-field theories do not contain fluctuations, and one finds indeed $c = 0$.

---

[7] To dissipate any belief that linear, massless dispersion relations would only belong to the fictitious worlds of the stringy theorist: exactly this kind of dispersion relation is actually realised in **graphene**, where the 'carriers of the charge behave as $(2+1)D$ ultra-relativistic particules without mass' [32].

1.3 The Virasoro Algebra

We now have to re-analyse the algebraic content of conformal invariance with a non-vanishing central charge $c \neq 0$. This can be done by a formal mode decomposition of the energy-momentum tensor $T(z) = \sum_{n=-\infty}^{\infty} L_n z^{-n-2}$. Alternatively, one may define the modes through their action on a primary scaling operator ϕ

$$L_n \phi(z_1, \bar{z}_1) = \frac{1}{2\pi i} \oint_{C_1} dz\, (z - z_1)^{n+1} T(z) \phi(z_1, \bar{z}_1) \tag{1.47}$$

where C_1 is a contour which encloses the point z_1. The complete set of the L_n (and similarly the $\bar{L}_n$ defined analogously) generates the complete set of $2D$ conformal transformations. From this definition, the commutators of the $L_n, \bar{L}_n$ can be worked out. Since we shall present the technique to do so when discussing later the free boson, we merely quote the result

$$[L_n, L_m] = (n - m)L_{n+m} + \frac{c}{12}(n^3 - n)\delta_{n+m,0}$$

$$[L_n, \bar{L}_m] = 0 \tag{1.48}$$

$$[\bar{L}_n, \bar{L}_m] = (n - m)\bar{L}_{n+m} + \frac{c}{12}(n^3 - n)\delta_{n+m,0}.$$

We see that the conformal generators form a pair of commuting **Virasoro algebra**s $\mathfrak{vir} \oplus \mathfrak{vir}$. Their structure is characterised by the value of the central charge c. The projective conformal transformations $\mathfrak{sl}(2, \mathbb{R}) \subset \mathfrak{vir}$ make up the maximal finite-dimensional sub-algebra and do not depend on c.

The novice might find it helpful to recall the algebraic theory of angular momentum: the Lie algebra $\mathfrak{so}(3) = \langle J_\pm, J_3 \rangle$ has in the chosen basis the commutators

$$[J_3, J_\pm] = \pm J_\pm, \qquad [J_+, J_-] = 2J_3. \tag{1.49}$$

The states of the representation R_j, denoted by $|jm\rangle$, are characterised by the quantum numbers j and m. The generator J_3 is used to measure m, via $J_3|jm\rangle = m|jm\rangle$. A **highest-weight state** is defined by the condition $J_+|jj\rangle = 0$. Then one uses a **ladder operator** to move between states, according to $J_-|jm\rangle = \alpha_{jm}|j\,m-1\rangle$ and the constant α_{jm} can be found from the commutator. Since $\mathfrak{so}(3)$ is a compact Lie algebra, its unitary representations R_j are finite-dimensional (indeed, if j is integer or half-integer, one can show that $J_-|j-j\rangle = 0$ which leads to dim $R_j = 2j + 1$).

We now outline the main facts of the representation theory of $\mathfrak{vir}$. In order to set up the algebraic machinery, reconsider the OPE with a primary operator (from now on, we drop the dependence on $\bar{z}_1$)

$$T(z)\phi(z_1) = \left(\frac{\Delta}{(z - z_1)^2} + \frac{1}{z - z_1}\frac{\partial}{\partial z_1}\right)\phi(z_1) + \text{regular terms}$$

$$= \sum_{n\in\mathbb{Z}} \frac{L_n(z_1)}{(z - z_1)^{n+2}}\phi(z_1) \tag{1.50}$$

and comparison gives the correspondences

$$L_{-1}(z_1)\phi(z_1) = \frac{\partial}{\partial z_1}\phi(z_1)$$
$$L_0(z_1)\phi(z_1) = \Delta\phi(z_1) \tag{1.51}$$
$$L_n(z_1)\phi(z_1) = 0 \quad \text{for all } n > 0.$$

Often, we shall write $L_n = L_n(0)$ for brevity. Denote the vacuum state by $|0\rangle$. *We postulate that $|0\rangle$ is conformally invariant*, i.e. $L_n|0\rangle = 0$ for all $n \geq -1$. Then define the **highest-weight state** as

$$|\Delta\rangle := \lim_{z\to 0} \phi(z)|0\rangle. \tag{1.52}$$

Then, *a scaling operator ϕ is primary if and only if* (i) $L_n|\Delta\rangle = 0$ *for all $n \geq 1$ and* (ii) $L_0|\Delta\rangle = \Delta|\Delta\rangle$. In constructing unitary representations of $\mathfrak{vir}$, one uses first the L_{-n} as **ladder operator**s to build up the states, according to

$$|\Delta; n_1, \ldots, n_k\rangle := L_{-n_k} \ldots L_{-n_1}|\Delta\rangle \tag{1.53}$$

with $n_1, \ldots, n_k > 0$.

Exercise 8 (i) A scaling operator ϕ is quasi-primary if and only if $L_0|\Delta\rangle = \Delta|\Delta\rangle$ and $L_1|\Delta\rangle = 0$. (ii) A scaling operator ϕ is primary if and only if $L_0|\Delta\rangle = \Delta|\Delta\rangle$ and $L_1|\Delta\rangle = L_2|\Delta\rangle = 0$.

The generator L_0 acts as a counting operator, according to

$$\begin{aligned} L_0|\Delta; n_1, \ldots, n_k\rangle &= L_0 L_{-n_k} \ldots L_{-n_1}|\Delta\rangle \\ &= n_k L_{-n_k} \ldots L_{-n_1}|\Delta\rangle + L_{-n_k} L_0 L_{-n_{k-1}} \ldots L_{-n_1}|\Delta\rangle \\ &= (n_k + \ldots + n_1 + \Delta)|\Delta; n_1, \ldots, n_k\rangle. \end{aligned} \tag{1.54}$$

The state $|\Delta; n_1, \ldots, n_k\rangle$ is called a **secondary state**, with **level** $n_1 + \ldots + n_k$. Any state which is not a highest-weight state related to a primary operator, according to (1.52), is a secondary state. We list a few examples of secondary states:

1. $|\Delta; 1, \ldots, 1\rangle = L_{-1}^n|\Delta\rangle = \lim_{z\to 0} \partial_z^n \phi(z)|0\rangle$.
2. $|T\rangle := L_{-2}|0\rangle = T(0)|0\rangle$ is a quasi-primary state, since $L_1|T\rangle = 0$. However, since $L_2|T\rangle = [L_2, L_{-2}]|0\rangle = c/2|0\rangle$, it is not primary for $c \neq 0$.
3. Consider the following state, at level $n = 2$

$$|\Phi^{(2)}\rangle := \left(L_{-2} - \frac{3}{2(2\Delta + 1)}L_{-1}^2\right)|\Delta\rangle. \tag{1.55}$$

Straightforward commutator calculations give

$$L_1|\Phi^{(2)}\rangle = 0, \qquad L_2|\Phi^{(2)}\rangle = \frac{1}{2}\frac{16\Delta^2 + (2c - 5)\Delta + c}{2\Delta + 1}|\Delta\rangle, \tag{1.56}$$

hence the state is quasi-primary. However, it is only primary, if a certain relation between c and Δ is satisfied. For example, if we take $c = \frac{1}{2}$, the state $|\Phi^{(2)}\rangle$ is primary only if $\Delta = \frac{1}{16}$ or $\frac{1}{2}$.

Pursuing the quest for unitary representations of $\mathfrak{vir}$, one must next define the notion of a *length*, which in turn requires the construction of **dual states** $\langle\Delta|$, so that one may naturally write for the norm $\|L_n|\Delta\rangle\|^2 = \langle\Delta|L_n^\dagger L_n|\Delta\rangle$. A natural way of defining such a structure is to work in polar coordinates, $z = \rho e^{i\varphi}$ and to introduce a **radial ordering**. Then a natural duality map relates $\rho \ll 1$ with $\rho \gg 1$ and we shall use here as a *duality map* the projective conformal transformation

$$z \mapsto z' = -\frac{1}{z} \tag{1.57}$$

and from (1.40) it follows that $z^2 T(z) = z'^2 T(z')$. Therefore

$$\sum_{n\in\mathbb{Z}} z^{-n} L_n = z^2 T(z) = z'^2 T(z') = \sum_{n\in\mathbb{Z}} z^{+n} L_n^\dagger \tag{1.58}$$

where one defines the adjoint generators $L_n^\dagger$ as the dual components. One may read off the **hermiticity condition**

$$L_n^\dagger = L_{-n}. \tag{1.59}$$

This is a natural result since the commutator $[L_n, L_{-n}]$ is in the Cartan sub-algebra $\mathfrak{h} = \langle L_0, c\rangle$ of $\mathfrak{vir}$. Especially, $L_0^\dagger = L_0$ is hermitian, so that the conformal weights Δ must be real numbers. In contrast to finite-dimensional Lie algebras (such as $\mathfrak{so}(3)$), the hermiticity of the generators is not sufficient to guarantee the unitarity of the representations, as we shall discuss below for the case of the Virasoro algebra.

The above discussion was centred at the origin but can be brought to an arbitrary position z_1 by the relation $\phi(z_1)|0\rangle = e^{z_1 L_{-1}}|\Delta\rangle$. Since under the duality transformation (1.57), a primary operator becomes $\phi(z') = z^{2\Delta}\phi(z)$, the dual states (located at infinity) are constructed as

$$\langle\Delta| := \lim_{z_1\to\infty} \langle 0|\phi(z_1) z_1^{2\Delta}$$

and one may check that the orthogonality relation $\langle\Delta|\Delta'\rangle = \delta_{\Delta,\Delta'}$ holds true. At a different position z_1, the dual state is $\langle 0|\phi(z_1) = \langle\Delta|z_1^{-2L_0} e^{(1/z_1)L_1}$.

Exercise 9 Verify that the two-point function $\langle 0|\phi(z_1)\phi(z_2)|0\rangle = (z_1 - z_2)^{-2\Delta} = \langle\phi(z_1)\phi(z_2)\rangle$ is correctly reproduced. The useful identity $\langle\Delta|L_1^n L_{-1}^m|\Delta\rangle = \delta_{n,m} n! \Gamma(n+2\Delta)/\Gamma(2\Delta)$ can be proven by induction, where $\Gamma(x)$ is Euler's Gamma-function.

We have already mentioned that the conformal transformation $w = \frac{L}{2\pi}\ln z$ transforms the complex plane $\mathbb{C}$ into the infinitely long strip of finite width L, and with periodic boundary conditions. Then, the quantum hamiltonian (energy operator) and the momentum operator on the strip can be expressed in terms of the modes $L_0, \bar{L}_0$ as follows:

$$H = \frac{2\pi}{L}(L_0 + \bar{L}_0) - \frac{\pi c}{6}\frac{1}{L}, \qquad P = \frac{2\pi}{L}(L_0 - \bar{L}_0). \tag{1.60}$$

We shall need these relations below several times, for instance when discussing modular invariance. Implicitly, the ground-state energy has been set to $E_0 = 0$ in the $L \to \infty$ limit.

Proof To derive (1.60), recall that the quantum hamiltonian H and the momentum operator P can be expressed in terms of the energy-momentum tensor $T(w)$ on the strip, with the coordinates $w = u + iv$

$$H = \frac{1}{2\pi} \int_0^L dv \left(T(w) + \bar{T}(\bar{w})\right), \qquad P = \frac{1}{2\pi} \int_0^L dv \left(T(w) - \bar{T}(\bar{w})\right).$$

From the transformation law (1.42), one has $T(w)\,dw = (\frac{2\pi}{L})^2 [z^2 T(z) - \frac{c}{24}]\,dw$ and combining this with the mode expansion (1.50) in the plane, it follows that

$$\frac{1}{2\pi} \int_0^L dv\, T(w) = \frac{2\pi}{L^2} \int_0^L dv \left[\sum_{n=-\infty}^{\infty} L_n \exp\left(-n\frac{2\pi}{L}(u+iv)\right) - \frac{c}{24}\right]$$

$$= \frac{2\pi}{L}\left[L_0 - \frac{c}{24}\right].$$

An analogous argument applies to $\bar{T}(\bar{w})$. Combining these gives the assertion. $\qquad\square$

1.4 Kac Formula and Unitary Minimal Models

Unitary representations are characterised by the absence of negative-norm states and we now inquire whether the hermiticity condition (1.59) of the Virasoro generators is sufficient to obtain it.

We carry out this discussion for the various levels of the representation. At level 0, no information is obtained, since we have simply the normalisation $\langle \Delta | \Delta \rangle = 1$. At level 1, we have the single state $L_{-1}|\Delta\rangle$ with norm

$$\left\| L_{-1}|\Delta\rangle \right\|^2 = \langle \Delta | L_1 L_{-1} | \Delta \rangle = 2\Delta. \tag{1.61}$$

Therefore, a necessary condition for unitarity is $\Delta \geq 0$. In the same way, since the norm of the energy-momentum tensor is

$$\left\| T|0\rangle \right\|^2 = \langle 0 | L_2 L_{-2} | 0 \rangle = \frac{c}{2}, \tag{1.62}$$

the further necessary condition $c \geq 0$ for unitarity follows.

Significant information is obtained at level 2. There are two independent states which we take to be $L_{-2}|\Delta\rangle$ and $L_{-1}^2|\Delta\rangle$. Unitarity requires that the matrix

$$\mathcal{M}_2 := \begin{pmatrix} \langle \Delta | L_2 L_{-2} | \Delta \rangle & \langle \Delta | L_2 L_{-1}^2 | \Delta \rangle \\ \langle \Delta | L_1^2 L_{-2} | \Delta \rangle & \langle \Delta | L_1^2 L_{-1}^2 | \Delta \rangle \end{pmatrix} \tag{1.63}$$

should be positive definite. A necessary condition for this is the positive definiteness of the **Kac determinant**

Fig. 1.4 Vanishing curves
$\Delta = \Delta_{r,s}$ for the first four
levels. In the *regions labelled*
according to their level n,
unitarity does not hold, since
$\det_n(c, \Delta) < 0$. The *first few*
intersection points are
indicated [26]

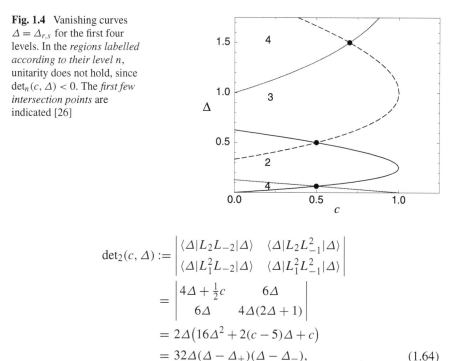

$$\det_2(c, \Delta) := \begin{vmatrix} \langle \Delta | L_2 L_{-2} | \Delta \rangle & \langle \Delta | L_2 L_{-1}^2 | \Delta \rangle \\ \langle \Delta | L_1^2 L_{-2} | \Delta \rangle & \langle \Delta | L_1^2 L_{-1}^2 | \Delta \rangle \end{vmatrix}$$

$$= \begin{vmatrix} 4\Delta + \tfrac{1}{2}c & 6\Delta \\ 6\Delta & 4\Delta(2\Delta + 1) \end{vmatrix}$$

$$= 2\Delta\big(16\Delta^2 + 2(c - 5)\Delta + c\big)$$

$$= 32\Delta(\Delta - \Delta_+)(\Delta - \Delta_-), \tag{1.64}$$

where we also wrote down its important factorisation property, and let

$$\Delta_\pm(c) := \frac{5 - c \pm \sqrt{(1 - c)(25 - c)}}{16}. \tag{1.65}$$

Hence $\det_2(c, \Delta)$ changes sign when passing in the (c, Δ) plane one of the curves
$\Delta = \Delta_\pm(c)$. One may define in the (c, Δ) plane a region (labelled '2' in Fig. 1.4),
bounded by the two curves $\Delta = \Delta_\pm(c)$, where the theory cannot be unitary. On
the other hand, in the outside of this region compatibility with unitarity is not yet
affected.

Very similar results hold true at any level n. Using the following parametrisation

$$c = c_m := 1 - \frac{6}{m(m + 1)}$$

$$\Delta_{r,s} = \Delta_{m-r,m+1-s} := \frac{[r(m + 1) - sm]^2 - 1}{4m(m + 1)}, \tag{1.66}$$

the Kac determinant is given by the celebrate **Kac formula**

$$\det_n(c, \Delta) = a_n \prod_{r,s=1;\, 1 \leq rs \leq n}^{n} (\Delta - \Delta_{r,s})^{p(n-rs)}, \tag{1.67}$$

where a_n is a known positive constant, r and s are positive integers and $p(k)$ is the
number of partitions of the integer k.

As for level 2, one may identify regions where the Kac determinant is negative
and unitarity cannot be satisfied. In Fig. 1.4, the non-unitary regions are indicated for

the levels 3 and 4. This makes it plausible, and indeed can be formally demonstrated, that the entire region $\Delta > 0$ and $0 < c < 1$ will be non-unitary. The only exception to this are certain *intersection points* between several of the lines $\Delta = \Delta_{r,s}$, the first few of which are also indicated in Fig. 1.4.

At level 2, we have already written down the state $|\Phi^{(2)}\rangle$ which indeed becomes primary along the two curves $\Delta = \Delta_{1,2} = \Delta_-(c)$ and $\Delta = \Delta_{2,1} = \Delta_+(c)$. Furthermore, its norm vanishes, $\||\Phi^{(2)}\rangle\| = 0$. One calls $|\Phi^{(2)}\rangle$ a **null state** and the corresponding primary scaling operator a **null operator**. A unitary representation can only be obtained if the quotient space with respect to $|\Phi^{(2)}\rangle$ is considered. Remarkably, and this holds true since the Kac determinant $\det_n(c, \Delta)$ always divides $\det_{n+1}(c, \Delta)$, analogous conclusions hold true at any level n. One identifies the set of null states and considers the quotient space with respect to them. If the parameter m in the parametrisation (1.66) of the central charge is an integer $m = 2, 3, 4, 5, \ldots$, it can be mathematically proven that no non-positive norm states remain in the resulting quotient space, but the details go much beyond the scope of this introduction.

The primary operators $\phi_{r,s}$, with conformal weight $\Delta = \Delta_{r,s}$, so identified, make up what is called a **unitary minimal model**. These are the most simple possible conformal field-theories. For each given integer $m \geq 2$, a unitary minimal model contains a finite list of admissible primary operators $\phi_{r,s}$, with $1 \leq r \leq m - 1$ and $1 \leq s \leq m$. Their conformal weights $\Delta_{r,s}$ are listed in the **Kac table**. Summarising:

Theorem (Friedan, Qiu, Shenker) *The Virasoro algebra admits unitary representations in exactly two cases*:

1. $c \geq 1$ *and* $\Delta \geq 0$, *with* a priori *an infinite number of primary operators.*
2. *The unitary minimal models, characterised by an integer* $m = 2, 3, 4, \ldots$ *The central charge* $c = c_m < 1$ *is given by* (1.66) *and the Kac table gives the finite list of the conformal weights* $\Delta_{r,s}$, *with* $1 \leq r \leq m - 1$ *and* $1 \leq s \leq m$.

While this theorem precisely states the necessary conditions for unitarity of the irreducible representations of $\mathfrak{vir}$, Goddard, Kent and Olive have shown, via the so-called '*coset construction*' that there is at least one explicitly unitary conformal field-theory for each unitary minimal model. The above conditions are therefore also sufficient for unitarity, at least for $c < 1$.

Example We illustrate the content of the theorem in the special case of the $2D$ Ising universality class. We had already mentioned before that $c = \frac{1}{2}$ in this universality class, which corresponds to $m = 3$. The Kac table then takes the following form

r	s		
	1	2	3
2	$\frac{1}{2}$	$\frac{1}{16}$	0
1	0	$\frac{1}{16}$	$\frac{1}{2}$

and we see that because of the symmetry $\Delta_{r,s} = \Delta_{m-r,m+1-s}$, each admissible value of the conformal weights occurs twice. Besides the identity operator **1**, with the conformal weight $\Delta_1 = 0$, there are two more primary operators. While the entire discussion focussed on the left-hand Virasoro generators L_n, it can be repeated identically for the right-handed generators $\bar{L}_n$. In trying to identify the primary operators with the physical observables in the Ising model, recall that the most interesting ones, such a magnetisation density and energy density, are *scalars* under rotations so that we can expect that $\Delta_{r,s} = \bar{\Delta}_{r,s}$. If one identifies

$$\mathbf{1} = \phi_{1,1} = \phi_{2,3}, \qquad \sigma = \phi_{1,2} = \phi_{2,2}, \qquad \varepsilon = \phi_{2,1} = \phi_{1,3} \qquad (1.68)$$

one obtains the scaling dimensions

$$x_1 = 2\Delta_{1,1} = 0, \qquad x_\sigma = 2\Delta_{1,2} = \frac{1}{8}, \qquad x_\varepsilon = 2\Delta_{2,1} = 1 \qquad (1.69)$$

which *correctly reproduce the known exact values of the 2D Ising model!* From (1.6), one may derive the values of the conventional critical 2D Ising model exponents $\alpha = 0$, $\beta = \frac{1}{8}$, $\nu = 1$ and so on.[8]

Exercise 10   The three-states **Potts model** may be described in terms of a discrete angular variable $\theta = 0, \frac{2\pi}{3}, \frac{4\pi}{3}$, and a classical hamiltonian $\mathcal{H} = -J \sum_{(i,j)} \cos(\theta_i - \theta_j)$, where the sum extends over the nearest-neighbour pairs of a square lattice. At the critical point, numerical estimates suggest with a high degree of accuracy that $c = \frac{4}{5}$ exactly. Can you reproduce the exact conventional critical exponents $\alpha = \frac{1}{3}$, $\beta = \frac{1}{9}$ and $\nu = \frac{5}{6}$ from a unitary minimal model? (Hint: the primary operators $\phi_{r,s}$ with $s = 2t$ even are *not* realised in the three-states Potts model.)

1.5 From Characters to Modular Invariance

Minimal models contain more than merely information on primary operators. A systematic overview on the secondary operators present is obtained through the **character**s of the irreducible representation of $\mathfrak{vir}$. In the generic case, that is not yet for minimal models, the character is for $\Delta \neq 0$

$$\chi(\Delta, c) = \chi_\Delta(q) := \mathrm{tr}\, q^{L_0 - c/24} = q^{\Delta - c/24} \sum_{n=0}^{\infty} p(n) q^n \qquad (1.70)$$

where again $p(n)$ stands for the number of **partitions** of the integer n and counts the number of secondary states at level n. We shall use below one or the other

---

[8]The physical content of a mathematical classification has still to be established by external evidence. To quote a well-known example, the periodic system of the chemical elements follows from the representation theory of the rotation Lie group $\mathfrak{so}(3)$. Still, that classification alone does not tell you that the 8th element keeps fires burning and allows vertebrates to breathe or that the 79th element has since prehistoric times attracted the greed of many.

notation, depending on whether the dependence on Δ and c or rather on q needs to be emphasised.

By definition, $p(0) = p(1) = 1$. Since $2 = 1 + 1$, one has $p(2) = 2$. Since $3 = 2 + 1 = 1 + 1 + 1$, one has $p(3) = 3$. Since $4 = 3 + 1 = 2 + 2 = 2 + 1 + 1 = 1 + 1 + 1 + 1$, one has $p(4) = 5$ and so on. A generating function is given by

$$\frac{1}{P(q)} := \sum_{n=0}^{\infty} p(n)q^n = \prod_{\ell=1}^{\infty}(1 - q^\ell)^{-1} \tag{1.71}$$

and long lists of values of $p(n)$ can be found in mathematical tables.

For unitary minimal models, the presence of null states requires a careful counting of the levels. For example, the primary operator $\phi_{r,s}$ has a null vector at level rs with conformal weight $\Delta_{r,s} + rs = \Delta_{m+r,m+1-s}$. Furthermore, since $\Delta_{r,s} = \Delta_{m-r,m+1-s}$, there is a further null state at level $(m-r)(m+1-s)$, with conformal weight $\Delta_{r,s} + (m-r)(m+1-s) = \Delta_{2m-r,s}$. These two primary states and their entire sets of secondary states must eliminated in the counting of secondary states of the unitary minimal character $\chi_{r,s}$. The arguments just given suggest that to this order, one has

$$\chi(\Delta_{r,s}, c_m) - \chi(\Delta_{2m-r,s}, c_m) - \chi(\Delta_{m+r,m+1-s}, c_m).$$

Continuing in this way, one obtains further null states at higher levels, correcting the last two contributions in the above expression. Remarkably, it turns out that the sets of higher null states intersect and that essentially two infinite ladders of null states exist. A careful analysis leads to the **Rocha-Caridi formula** for the unitary minimal character, with $P(q)$ having been defined in (1.71)

$$\begin{aligned}
\chi_{r,s} &= \chi(\Delta_{r,s}, c) - \sum_{k=1}^{\infty}\big[\chi(\Delta_{2km-r,s}, c) + \chi(\Delta_{r+m(2k-1),m+1-s}, c) \\
&\quad - \chi(\Delta_{(2k+1)m-r,m+1-s}, c) - \chi(\Delta_{r+2km,s}, c)\big] \\
&= \sum_{k=-\infty}^{+\infty}\big[\chi(\Delta_{r+2km,s}, c) - \chi(\Delta_{r+m(2k-1),m+1-s}, c)\big] \\
&= q^{\Delta_{r,s}-c/24}\frac{1}{P(q)}\sum_{k=-\infty}^{\infty}\big(q^{\Delta_{2mk+r,s}} - q^{\Delta_{2mk+r,-s}}\big).
\end{aligned} \tag{1.72}$$

The character $\chi_{r,s} = \chi_{r,s}(q) = q^{\Delta_{r,s}-c/24}\sum_{n=0}^{\infty}d_n(\Delta_{r,s})q^n$ counts the number $d_n(\Delta_{r,s})$ of secondary states at level n, and for which tables are available in the literature. Since characters are generating functions for the entire set of states in the irreducible representations of the Virasoro algebra, they provide this complete information a single analytic expression (mathematically, these are related to Jacobi θ-functions).

Exercise 11 Take into account that $L_{-1}\mathbf{1} = 0$ and derive the generic character $\chi(0, c) = (1 - q)q^{-c/24}P(q)^{-1}$. Further show that the generating function of the number of *quasi*-primary states at each level is given by $\chi_{QP}(\Delta, c) = (1 - q) \times \chi(\Delta, c) + q\delta_{\Delta,0}$.

In this way, the unitary representations of a single Virasoro algebra have been constructed. The representation space, built from the highest weight primary state with conformal weight $\Delta = \Delta_{r,s}$, is referred to as the **Verma module** and denoted generically by $\mathcal{V}_\Delta$, or simply by $\mathcal{V}_{r,s}$ for minimal models. The full Hilbert space of the conformal field-theory (CFT) is then made out of direct sums of pairs of Verma modules

$$\mathcal{V} = \bigoplus_{\Delta,\overline{\Delta}} n_{\Delta,\overline{\Delta}} \mathcal{V}_\Delta \otimes \mathcal{V}_{\overline{\Delta}} \tag{1.73}$$

where the positive integers $n_{\Delta,\overline{\Delta}}$ specify how many distinct primary operators with the conformal weights $(\Delta, \overline{\Delta})$ are present in the CFT under consideration. Therefore, a CFT (or a universality class in statistical mechanics) can be characterised by the value of the central charge c and the primary operators present in the theory. In a complete CFT, the primary operators satisfy between themselves an **operator product algebra** (OPA), given by the following operator product expansions (OPE) between the primary operators

$$\phi_a(z_1,\bar{z}_1)\phi_b(z_2,\bar{z}_2) = \sum_c \mathcal{C}_{abc} z_{12}^{-\Delta_{ab,c}} \bar{z}_{12}^{-\overline{\Delta}_{ab,c}} \phi_c(z_2,\bar{z}_2) + \text{regular terms} \tag{1.74}$$

where $z_{12} = z_1 - z_2$, $\Delta_{ab,c} := \Delta_a + \Delta_b - \Delta_c$ and similarly for the right-handed factor. This kind of expansion is supposed to be valid in the limit where $z_{12}, \bar{z}_{12} \to 0$ and when inserted into correlation functions. The OPE-coefficient $\mathcal{C}_{abc}$ was already met in Eq. (1.17), giving the projectively invariant three-point function. Explicit, if very cumbersome, formulæ exist which express $\mathcal{C}_{abc}$ through the central charge and the conformal weights.

The requirement of consistency of the OPA Eq. (1.74) with the existence of null operators is taken into account by the existence of **fusion algebras**. Written formally as $\mathcal{V}_a \odot \mathcal{V}_b = \sum_c N_{ab}^c \mathcal{V}_c$, this associative and commutative algebra describes which primary operators ϕ_c arise in the fusion of the primary operators ϕ_a and ϕ_b. For unitary minimal models of central charge $c = c_m$, one has explicitly

$$V_{r_1,s_1} \odot V_{r_2,s_2} = \sum_{r_3=|r_1-r_2|}^{r_1+r_2-1}{}' \sum_{s_3=|s_1-s_2|}^{s_1+s_2-1}{}' V_{r_3,s_3} \tag{1.75}$$

and the primes on the sum indicate the restrictions $1 \le r_3 \le m-1$ and $1 \le s_3 \le m$.

Exercise 12 Check that in the $2D$ Ising universality class, one has the fusion algebra

$$\sigma \odot \sigma = 1 + \varepsilon, \qquad \varepsilon \odot \varepsilon = 1, \qquad \sigma \odot \varepsilon = \sigma$$

and obviously, $1 \odot 1 = 1$, $1 \odot \sigma = \sigma$ and $1 \odot \varepsilon = \varepsilon$. Is this compatible with the global $\mathbb{Z}_2$-symmetry of the classical Ising hamiltonian?

While the fusion algebras limit the admissible values of $(\Delta_{r,s}, \overline{\Delta}_{r,s})$, they do not tell which ones are actually realised. This last piece of information comes from the

Fig. 1.5 Parametrisation of
the torus by $\tau \in \mathbb{C}$ and its
modular transformation
generated by (**a**) T and
(**b**) S [26]

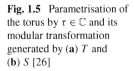

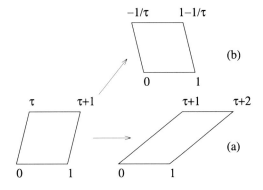

Fig. 1.5 Parametrisation of the torus by $\tau \in \mathbb{C}$ and its modular transformation generated by (**a**) T and (**b**) S [26]

requirement of **modular invariance** on the torus. Mathematically, a **torus** is given
by a parallelogramme in the complex plane, with vertices at the points $0, 1, \tau, 1 + \tau$
and with the opposite edges identified, see Fig. 1.5. Starting from a cylinder with
unit circumference and of length $\mathrm{Im}\,\tau$, by twisting the ends by a relative amount $\mathrm{Re}\,\tau$
and glueing them together. On the cylinder the energy and momentum operator of a
CFT can be written as $H = 2\pi(L_0 + \bar{L}_0) - \pi c/6$ and $P = 2\pi(L_0 - \bar{L}_0)$. Therefore,
the partition function of the CFT on the torus reads

$$Z = \mathrm{tr}\,e^{-2\pi\,\mathrm{Im}\,\tau H + 2\pi i\,\mathrm{Re}\,\tau P} = \mathrm{tr}\,q^{L_0 - c/24}\bar{q}^{\bar{L}_0 - c/24} = \sum_{\Delta,\bar{\Delta}} n_{\Delta,\bar{\Delta}}\chi_\Delta(q)\chi_{\bar{\Delta}}(\bar{q}) \quad (1.76)$$

where we have set $q := e^{2\pi i \tau}$. It is a well-known mathematical fact that the
parametrisation of the torus introduced above is not unique. For example, the transformations $S : \tau \mapsto -1/\tau$ and $T : \tau \mapsto \tau + 1$ give the same torus, see Fig. 1.5.
Indeed, they generate together the **modular group** $Sl(2, \mathbb{Z})$. In general, **modular
transformation**s are of the form $\tau \mapsto \tau' = \frac{a\tau + b}{c\tau + d}$ with $ad - bc = 1$ and a, b, c, d
being *integer*. Since the modular transformation always give back the *same* torus,
the partition function Z should be modular invariant. For unitary minimal models,
the modular transformations are given by the change $q \mapsto \tilde{q} := \exp(-2\pi i/\tau)$ and

$$
\begin{aligned}
T: &\quad \chi_{r,s}(q) \to \exp\left[2\pi i\left(\Delta_{r,s} - \frac{c}{24}\right)\right]\chi_{r,s}(q) \\
S: &\quad \chi_{r,s}(q) \to \sum_{r',s'} S_{r,s}^{r',s'}\chi_{r',s'}(q),
\end{aligned}
\quad (1.77)
$$

where

$$S_{r,s}^{r',s'} = \sqrt{\frac{8}{m(m+1)}}(-1)^{(r+s)(r'+s')}\sin\left(\frac{\pi r r'}{m}\right)\sin\left(\frac{\pi s s'}{m+1}\right). \quad (1.78)$$

Invariance under T simply means that $\Delta_{r,s} - \bar{\Delta}_{r,s}$ must be an integer, but the constraints following from the invariance under S are highly non-trivial. The resulting
invariant partition functions have been classified by Cappelli, Itzykson and Zuber
and it turns out that there is a close relationship to Cartan's classification of the

simple complex Lie algebras ADE. The most simple solution is the **diagonal** one $n_{\Delta,\bar{\Delta}} = \delta_{\Delta,\bar{\Delta}}$, which gives the partition function of the so-called 'A-series'

$$Z^{(A)} = \frac{1}{2} \sum_{r=1}^{m-1} \sum_{s=1}^{m} |\chi_{r,s}|^2. \tag{1.79}$$

In writing down the **operator content** of a model, that is the list of primary operators and their secondary states, one sometimes also uses the notations $(\Delta_{r,s}) = \chi_{r,s}$ and $(\Delta_{r,s}, \bar{\Delta}_{r',s'}) = \chi_{r,s} \bar{\chi}_{r',s'}$ and indicates the vir-representations by the value of their conformal weights. For example, in this notation $Z^{(A)} = \frac{1}{2} \sum_{r=1}^{m-1} \sum_{s=1}^{m} (\Delta_{r,s}, \Delta_{r,s})$ and it is clear that all primary operators therein are scalars.

Example For the $2D$ Ising universality class, a minimal model with $m = 3$, this gives $Z_{\text{Ising}} = |\chi_{1,1}|^2 + |\chi_{1,2}|^2 + |\chi_{2,1}|^2 = (0,0) + (\frac{1}{16}, \frac{1}{16}) + (\frac{1}{2}, \frac{1}{2})$. Put into words, all primary operators are predicted to be scalar and the complete list $\mathbf{1} = (0,0)$, $\sigma = (\frac{1}{16}, \frac{1}{16})$, $\varepsilon = (\frac{1}{2}, \frac{1}{2})$ of admissible primary operators is present, in agreement with the 'experimental' identifications we admitted earlier. Especially, the scaling dimensions $x_{\mathbf{1}} = 0$, $x_{\sigma} = 2\Delta_{1,2} = \frac{1}{8}$ and $x_{\varepsilon} = 2\Delta_{2,1} = 1$ are immediately read off.

Different fusion algebras will in general lead to different modular-invariant partition functions. For example, there are two modular-invariant partition functions for the unitary minimal model with $m = 5$, or $c = \frac{4}{5}$. The first one, which corresponds to the A-series, describes what is known as the 'tetracritical Ising model', but there is a second solution, which only contains the primary operators $\phi_{r,s}$ with $s = 1, 3, 5$ and which describes the universality class of the three-states Potts model, whose partition function is the first member of the '*D-series*'. Explicitly

$$Z_{\text{Potts-3}} = |\chi_{1,1}|^2 + |\chi_{2,1}|^2 + |\chi_{3,1}|^2 + |\chi_{4,1}|^2 + 2|\chi_{2,3}|^2 + 2|\chi_{1,3}|^2$$

$$+ \chi_{1,1}\bar{\chi}_{4,1} + \chi_{2,1}\bar{\chi}_{3,1} + \chi_{3,1}\bar{\chi}_{2,1} + \chi_{4,1}\bar{\chi}_{1,1}$$

$$= (0,0) + \left(\frac{2}{5}, \frac{2}{5}\right) + \left(\frac{7}{5}, \frac{7}{5}\right) + (3,3) + 2\left(\frac{1}{15}, \frac{1}{15}\right) + 2\left(\frac{2}{3}, \frac{2}{3}\right)$$

$$+ (0,3) + \left(\frac{2}{5}, \frac{7}{5}\right) + \left(\frac{7}{5}, \frac{2}{5}\right) + (3,0). \tag{1.80}$$

One may read off the scaling dimensions $x_{\sigma} = 2\Delta_{2,3} = \frac{2}{15}$ and $x_{\varepsilon} = 2\Delta_{2,1} = \frac{4}{5}$.

Still, the ADE-classification of the modular-invariant partition functions does *not* amount to a full classification of the $2D$ conformally invariant universality classes, not even in the restricted context of minimal models. Examples of distinct unitary minimal models, distinguished by their OPA, but with the *same* modular-invariant partition functions, are known. The first one occurs in the A-series with $m = 4$ [1].

Example We illustrate the modular transformations generated by S in a simple example. Let $\tau = i\delta$ with δ real, so that the torus becomes a rectangle. Then under

S, one has $\delta \to 1/\delta$ and $q \to \tilde{q} = e^{-2\pi/\delta}$. Consider the generating function of the partitions of the integers $P(q)$ Eq. (1.71)

$$P(q) = \prod_{n=1}^{\infty} \left(1 - e^{-2\pi\delta n}\right) = \sum_{n=-\infty}^{\infty} \exp\left(-\pi\delta\left(3n^2 + n\right) + i\pi n\right) \qquad (1.81)$$

by **Euler's pentagonal theorem**. The modular transformation is carried out with the help of the **Poisson re-summation formula**

$$\sum_{n=-\infty}^{\infty} f(n) = \sum_{p=-\infty}^{\infty} \int_{-\infty}^{\infty} dx \, f(x)e^{2\pi i p x}. \qquad (1.82)$$

Since $\cos((2p+1)\pi/6) = (-1)^n \frac{\sqrt{3}}{2}$ if $p = 3n$ or $p = 3n - 1$ and vanishes if $p = 3n + 1$, one finds

$$P(q) = \frac{2}{\sqrt{3\delta}} \exp\left[\frac{\pi}{12}\left(\delta - \delta^{-1}\right)\right] \sum_{p=-\infty}^{\infty} \cos\left(\frac{(2p+1)\pi}{6}\right) \exp\left(-\frac{p(p+1)\pi}{3\delta}\right)$$

$$= (-i\tau)^{-1/2} \exp\left[-\frac{i\pi}{12}\left(\tau + \frac{1}{\tau}\right)\right] P(\tilde{q}). \qquad (1.83)$$

As an application, we consider the field-theory of the *free boson*. Since the central charge of the $1D$ free boson is $c = 1$, one expects the generic character $q^{-1/24} P(q)^{-1}$. In view of the modular transformation properties of $P(q)$, it turns out, however, that the modular-invariant partition function of the free boson reads

$$Z_{\text{boson}} := q^{-1/24} \tilde{q}^{-1/24} \frac{1}{P(q)} \frac{1}{P(\tilde{q})} (\text{Im}\,\tau)^{-1/2}. \qquad (1.84)$$

The non-trivial factor $(\text{Im}\,\tau)^{-1/2}$ is usually obtained from a long and intricate discussion of the zero-mode subtraction in the functional integral. We shall come back to this result below.

We close this long section with two final observations:

(A) The null operators lead to very explicit predictions. We have seen that they lead to null states of vanishing norm. Therefore, when one inserts them into a correlator with other primary operators ϕ_a, one has $\langle \phi_1 \dots \phi_n \Phi^{(2)} \rangle = 0$, where for definiteness we used the explicit operator known at level 2. This leads to new, non-trivial differential equations for the n-point functions from which these can be found. For the above example, one finds

$$\left[\sum_{k=1}^{n}\left(\frac{\Delta_k}{(z - z_k)^2} + \frac{1}{z - z_k}\frac{\partial}{\partial z_k}\right) - \frac{3}{2(2\Delta + 1)}\frac{\partial^2}{\partial z^2}\right]$$

$$\times \langle \phi_1(z_1) \dots \phi_n(z_n)\phi(z) \rangle = 0. \qquad (1.85)$$

In this way, the use of unitary minimal models permits to solve a critical system from statistical mechanics. We refer to the literature for details.

(B) While unitarity is an essential ingredient in the construction of quantum field-theories, it is by no means a requirement in many of the models studied in statistical mechanics. Also, the Kac formula remains valid for non-integer values of m. For example, one may set $m = p/(p - p')$, where p and p' are relatively prime. Then one obtains a set of **non-unitary minimal models**, where Eq. (1.66) is replaced by

$$c = c_{p,p'} := 1 - \frac{6(p - p')}{pp'}$$
$$\Delta_{r,s} = \Delta_{p-r,p'-s} := \frac{(rp' - sp)^2 - (p - p')}{4pp'}, \tag{1.86}$$

and the Kac table lists the values of $\Delta_{r,s}$ with $1 \le r \le p - 1$ and $1 \le s \le p' - 1$. Since the conformal weights are rational numbers, this class of minimal models is also called **rational conformal field-theory** (RCFT). Symbolically, they are denoted by $\mathcal{M}_{p,p'}$ (we use here the convention $p < p'$) and the unitary minimal model is the special case $\mathcal{M}_{m,m+1}$ with $m = 2, 3, 4, \ldots$ an integer. The fusion algebra (1.75) remains valid, but where now the restrictions on the sum mean $1 \le r_3 \le p - 1$ and $1 \le s_3 \le p' - 1$. The Rocha-Caridi formula Eq. (1.72) can be generalised similarly, by replacing systematically $m \mapsto p$ and $m + 1 \mapsto p'$, and reads

$$\chi_{r,s} = \sum_{k=-\infty}^{\infty} \big(\chi(\Delta_{r+2kp,s}, c) - \chi(\Delta_{r+p(2k-1), p'-s}, c) \big). \tag{1.87}$$

The most simple non-unitary minimal model is found for $p = 2$ and $p' = 5$. Then $c = -\frac{22}{5}$ and the Kac table contains two primary operators with the conformal weights $\Delta_{1,1} = 0$ and $\Delta_{1,2} = -\frac{1}{5}$. This conformal theory describes the scaling properties of the **Yang-Lee singularity** of the critical $2D$ Ising model in an *imaginary* magnetic field. Another example of a non-unitary theory is **percolation**, where the central charge $c = 0$. This model will be discussed in depth in later chapters of this volume.

1.6 The Free Boson

In order to illustrate and make more concrete the abstract developments in the previous sections, we shall now discuss in detail the conformal field-theory of the free boson. In complex coordinates, the action is $S = \int d\bar{z} \wedge dz\, \partial_z \varphi \partial_{\bar{z}} \varphi$ which gives the classical equation of motion $\partial_z \partial_{\bar{z}} \varphi = 0$. Then one would write the general solution as the sums of two terms: $\varphi(z, \bar{z}) = \varphi(z) + \bar{\varphi}(\bar{z})$. Similarly, the associated Green's function $G(z, \bar{z}) := \langle \phi(z, \bar{z})\varphi(0, 0) \rangle$ satisfies the equation $\partial_z \partial_{\bar{z}} G(z, \bar{z}) = \delta^{(2)}(z, \bar{z})$, with the formal solution

$$G(z, \bar{z}) = -4 \ln \frac{|z|}{R} = -2 \ln \frac{z}{R} - 2 \ln \frac{\bar{z}}{R}$$

where R is a constant. This is nothing but Coulomb's law in two dimension, so that the system is often called the **Coulomb gas**. However, the given solution is problematic for the construction of a conformal field-theory, since if R is finite, it does break scale-invariance and for $R \to \infty$, the Green's function is ill-defined.

Rather than trying to remedy this, we shall concentrate on a different kind of observables which do not present this difficulty. Consider the **currents**

$$J := \frac{i}{2} \partial_z \varphi, \qquad \bar{J} := -\frac{i}{2} \partial_{\bar{z}} \varphi. \tag{1.88}$$

The equations of motion are the conservation laws $\partial_{\bar{z}} J = \partial_z \bar{J} = 0$. Hence $J = J(z)$ and $\bar{J} = \bar{J}(\bar{z})$ are analytic or anti-analytic, respectively. These currents have well-defined two-point functions

$$\langle J(z_1) J(z_2) \rangle = \frac{1/2}{(z_1 - z_2)^2}$$

$$\langle \bar{J}(\bar{z}_1) \bar{J}(\bar{z}_2) \rangle = \frac{1/2}{(\bar{z}_1 - \bar{z}_2)^2} \tag{1.89}$$

$$\langle J(z_1) \bar{J}(\bar{z}_2) \rangle = 0$$

as one may formally check by symbolic derivation of the Green's function result mentioned above. We shall use Eqs. (1.89) as the defining properties of the free boson and shall make them the starting point of the construction of the associated CFT. One reads off the scaling dimensions $x_J = x_{\bar{J}} = 1$ and spins $s_J = 1 = -s_{\bar{J}}$.

The quantisation of this classical theory requires to turn the currents $J, \bar{J}$ into operators. As operators, they will in general *not* commute at different positions, $[J(z), J(z')] \neq 0$! The necessary ordering in products of such operators is achieved here by a (bosonic free-field) **radial ordering** such that for any two space-dependent operators $A(z)$ and $B(w)$ one has

$$\mathcal{R}\big(A(z) B(w)\big) := \begin{cases} A(z) B(w); & \text{if } |z| > |w| \\ B(w) A(z); & \text{if } |z| < |w|. \end{cases} \tag{1.90}$$

Returning to the free boson, the classical averages (1.89) suggest that a sensible choice for the OPE of two currents should be

$$\mathcal{R}\big(J(z) J(z')\big) = \frac{1/2}{(z - z')^2} + \text{regular terms}$$

$$\mathcal{R}\big(J(z) \bar{J}(\bar{z}')\big) = \text{regular terms} \tag{1.91}$$

$$\mathcal{R}\big(\bar{J}(\bar{z}) \bar{J}(\bar{z}')\big) = \frac{1/2}{(\bar{z} - \bar{z}')^2} + \text{regular terms}.$$

As usual, such an OPE will acquire a meaning when inserted into a correlation function and one concentrates on the singular behaviour when $z - z' \to 0$.

The next step is to write a meromorphic mode expansion of the currents

$$J(z) = \sum_{n \in \mathbb{Z}} J_n z^{-n-1}, \qquad \bar{J}(\bar{z}) = \sum_{n \in \mathbb{Z}} \bar{J}_n \bar{z}^{-n-1} \tag{1.92}$$

or equivalently the modes can be written in terms of a contour integral

$$J_n = \oint_{C_0} \frac{dz}{2\pi i} J(z) z^n, \qquad \bar{J}_n = \oint_{\bar{C}_0} \frac{d\bar{z}}{2\pi i} \bar{J}(\bar{z}) \bar{z}^{:n} \tag{1.93}$$

Fig. 1.6 Re-organisation of complex integration contours

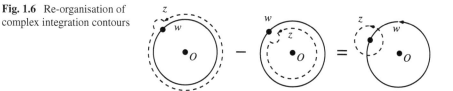

and where $C_0, \bar{C}_0$ are simple contours around the origin. The OPE (1.91) is equivalent to the following set of commutator relations

$$[J_n, J_m] = \frac{n}{2}\delta_{n+m,0}, \qquad [J_n, \bar{J}_m] = 0, \qquad [\bar{J}_n, \bar{J}_m] = \frac{n}{2}\delta_{n+m,0} \qquad (1.94)$$

which form a $\mathfrak{u}(1) \oplus \mathfrak{u}(1)$ **Kac-Moody algebra**.

Proof The first one of the commutators (1.94) is proven as follows. Form the commutator

$$[J_n, J_m]$$
$$= \oint_{C_0} \frac{dz}{2\pi i} z^n \oint_{C_0} \frac{dw}{2\pi i} w^m J(z)J(w) - \oint_{C_0} \frac{dw}{2\pi i} w^m \oint_{C_0} \frac{dz}{2\pi i} z^n J(w)J(n)$$
$$= \oint_{C_0} \frac{dw}{2\pi i} w^m \left[\oint_{|z|>|w|} \frac{dz}{2\pi i} z^n - \oint_{|z|<|w|} \frac{dz}{2\pi i} z^n \right] \mathscr{R}\big(J(z)J(w)\big)$$
$$= \oint_{C_0} \frac{dw}{2\pi i} w^m \oint_{C_w} \frac{dz}{2\pi i} z^n \mathscr{R}\big(J(z)J(w)\big)$$
$$= \oint_{C_0} \frac{dw}{2\pi i} w^m \oint_{C_w} \frac{dz}{2\pi i} z^n \left(\frac{1/2}{(z-w)^2} + \text{regular terms} \right)$$
$$= \oint_{C_0} \frac{dw}{2\pi i} w^m \frac{1}{2} n w^{n-1} = \frac{n}{2}\delta_{n+m,0}$$

where C_0 and C_w are simple contours (circles) around 0 and w, respectively. In the third line, the radial ordering was introduced and in the forth line, reference was made to Fig. 1.6 for the re-writing of the integration contours for the variables z (dashed lines) and w (full lines). In the fifth line, recall that the 'regular terms' are regular for $z - w \to 0$ and hence do not contribute to the contour integral, whose singular contributions are found from the residue theorem. The other two commutators (1.94) are derived similarly.

The derivation of the Virasoro algebra from the OPE of $T(z)$ with itself, see p. 15, is done analogously. □

Kac-Moody algebras can be constructed for more general Lie algebras. Consider a finite-dimensional Lie algebra $\mathfrak{g}$, whose generators J^a satisfy the commutator relations $[J^a, J^b] = f^{ab}{}_c J^c$. If these generators can be associated to conserved currents $J^a(z)$ such that the following OPE is valid

$$J^a(z)J^b(w) = \frac{k}{(z-w)^2}\delta^{ab} + f^{ab}{}_c \frac{1}{z-w} J^c(w) + \text{regular terms} \qquad (1.95)$$

the modes J_n^a, defined according to $J^a(z) = \sum_{n\in\mathbb{Z}} J_n^a z^{-n-1}$ satisfy the commutation relations of the $\mathfrak{g}$-Kac-Moody algebra (usually denoted by $\widehat{\mathfrak{g}}_k$)

$$[J_n^a, J_m^b] = f^{ab}{}_c J_{n+m}^c + kn\delta^{ab}\delta_{n+m,0}. \tag{1.96}$$

Particle physicists had discussed the algebras under the name of **current algebras** long before they became of interest in statistical mechanics and condensed matter, and before their extensive study by mathematicians, under the name of '*affine Lie algebras*'. The constant k is called the **central charge** (or the **level**) of the Kac-Moody algebra, whose value is characteristic for the Kac-Moody algebra, and characterises the importance of the quantum fluctuations. For non-abelian Lie algebras $\mathfrak{g}$, the corresponding fields are no longer free but rather reflect the presence of topological terms in the so-called '*Wess-Zumino-Witten models*'.

Exercise 13 Use the mode expansion of a primary operator $\phi(z) = \sum_{m\in\mathbb{Z}} \phi_m \times z^{-m-\Delta}$ of conformal weight Δ, and the OPE equation (1.50) with the energy-momentum tensor $T(z) = \sum_{n\in\mathbb{Z}} L_n z^{-n-2}$ in order to derive the mode commutator

$$[L_n, \phi_m] = \big(n(\Delta - 1) - m\big)\phi_{n+m}.$$

1.7 The Sugawara Construction

The next step in constructing a CFT for the free boson consists in finding the energy-momentum tensor. A heuristic argument from particle physics can be used to motivate the final choice. Critical phenomena are interested in the long-distance, or low-energy, behaviour of the n-point functions. Particle physicists have the habit of looking at effective low-energy theories, where in the Feynman diagrams to be considered the propagator is simply replaced by an effective coupling constant.[9] Then the energy should be proportional to the product of the interacting currents. This suggests that a good candidate for an energy-momentum tensor might be of the form $T \sim :JJ:$. Of course, a naïve product of two currents would be ill-defined because of the singular current-current two-point function and a '*normal ordering*' procedure, indicated here by $::$, must be used to eliminate this difficulty. The solution consists of considering the product of two currents at a small distance ε ('*point-splitting procedure*') and letting $\varepsilon \to 0$ when appropriate.

With these heuristics in mind, we shall consider here the following definition

$$T(z) = :JJ:(z) := \lim_{\varepsilon\to 0}\left(J(z+\varepsilon)J(z-\varepsilon) - \frac{1}{8\varepsilon^2}\right) \tag{1.97}$$

(it can be shown that the precise form of the point-splitting is not important, but the symmetric form used here shortens the calculations). In order to justify this

---

[9]A good example of this is the Fermi theory of weak interactions, where the momentum-dependence of the propagators of the intermediate weak bosons $W^\pm$ and Z of the unified electroweak theory can be neglected for energies $\ll M_{W,Z} c^2 \approx 80$ [GeV].

definition, one must show that the operator $T(z)$ has the expected properties of an energy-momentum tensor of a conformal field-theory.

Obviously, one has $\langle T(z) \rangle = 0$, which explains the choice made in (1.97) for the constant term. Next, we consider the OPE of $T(z)$ with the current $J(w)$ and then with $T(w)$ itself, which gives

$$T(z)J(w) = \frac{1}{(z-w)^2} J(w) + \frac{1}{z-w} \partial_w J(w) + \text{regular terms}$$

$$T(z)T(w) = \frac{1}{2}\frac{1}{(z-w)^4} + \frac{2}{(z-w)^2} T(w) + \frac{1}{z-w} \partial_w T(w) + \text{regular terms}.$$

(1.98)

This is indeed the expected OPE for an energy-momentum tensor and one confirms once more the central charge $c = 1$ for the free boson. Furthermore, the conserved current $J(z)$ is a primary operator, with the expected conformal weight $\Delta_J = 1$.

Proof These OPEs can be checked as follows. First, consider

$$T(z')J(z)$$

$$= \lim_{\varepsilon \to 0} \left(J(z'+\varepsilon)J(z'-\varepsilon)J(z) - \frac{1}{8\varepsilon^2} J(z) \right)$$

$$= \lim_{\varepsilon \to 0} \left(\frac{1}{2(z'-z+\varepsilon)^2} J(z'-\varepsilon) + \frac{1}{2(z'-z-\varepsilon)^2} J(z'+\varepsilon) - \frac{1}{8\varepsilon^2} J(z) \right)$$

$$= \frac{1}{(z'-z)^2} J(z) + \frac{1}{z'-z} \partial_z J(z) + \text{regular terms}$$

as asserted, and $J(z') = J(z) + (z'-z)\partial_z J(z) + \ldots$ was used in the last step. The terms not explicitly spelled out are regular in the limit $z'-z \to 0$. This identity helps to shorten the computation of the OPE of $T(z)$ with itself

$$T(z')T(z)$$

$$= \lim_{\varepsilon \to 0} \left\{ T(z') \left(J(z+\varepsilon)J(z-\varepsilon) - \frac{1}{8\varepsilon^2} \right) \right\}$$

$$= \lim_{\varepsilon \to 0} \left\{ \frac{1}{(z'-z-\varepsilon)^2} \left[J(z+\varepsilon)J(z-\varepsilon) - \frac{1}{8\varepsilon^2} + \frac{1}{8\varepsilon^2} \right] \right.$$

$$+ \frac{1}{(z'-z+\varepsilon)^2} \left[J(z+\varepsilon)J(z-\varepsilon) - \frac{1}{8\varepsilon^2} + \frac{1}{8\varepsilon^2} \right]$$

$$+ \frac{1}{z'-z-\varepsilon} \left[\left(\partial_z J(z+\varepsilon) \right) J(z-\varepsilon) + J(z+\varepsilon) \left(\partial_z J(z-\varepsilon) \right) \right]$$

$$\left. + \left[\frac{1}{z'-z+\varepsilon} - \frac{1}{z'-z-\varepsilon} \right] J(z+\varepsilon) \left(\partial_z J(z-\varepsilon) \right) - \frac{1}{8\varepsilon^2} T(z') \right\}. \quad (1.99)$$

Because of $\langle J(z_1, \bar{z}_1) \partial_{z_2} J(z_2, \bar{z}_2) \rangle = (z_1 - z_2)^{-3}$, the last line in the above expansion is potentially singular in the $\varepsilon \to 0$ limit and must be treated carefully. We find, using again the normal ordering prescription (1.97)

$$T(z')T(z) = \lim_{\varepsilon \to 0} \left\{ \frac{2}{(z'-z)^2} T(z) + \frac{1}{4\varepsilon^2} \frac{1}{(z'-z)^2} + \frac{3}{4} \frac{1}{(z'-z)^4} \right.$$

$$\left. + \frac{1}{z'-z} \partial_z T(z) - \frac{\varepsilon}{4\varepsilon^3} \frac{1}{(z'-z)^2} - \frac{1}{4} \frac{1}{(z'-z)^4} \right\} + \text{regular terms}$$

$$= \frac{1}{2} \frac{1}{(z'-z)^4} + \frac{2}{(z'-z)^2} T(z) + \frac{1}{z'-z} \partial_z T(z) + \text{regular terms}$$

$$(1.100)$$

as claimed. The last terms in (1.100) come from the last line in (1.99). The terms which are singular for $\varepsilon \to 0$ as well as for $z' - z \to 0$ cancel out, as it should be for a well-defined theory. $\square$

Besides the currents, the so-called **vertex operator**s, formally given by

$$V_\alpha(z) = :\exp i\alpha\varphi(z): \qquad (1.101)$$

play an important role. Again, a normal-ordering procedure must be introduced to make this definition meaningful, but for our limited purposes, this will not be explicitly needed. One finds the following OPEs

$$J(z)V_\alpha(w) = \frac{\alpha}{z-w} V_\alpha(w) + \text{regular terms}$$

$$T(z)V_\alpha(z) = \frac{\alpha^2}{(z-w)^2} V_\alpha(w) + \frac{1}{z-w} \frac{\partial}{\partial w} V_\alpha(w) + \text{regular terms} \qquad (1.102)$$

$$V_\alpha(z)V_\beta(w) = \delta_{\alpha+\beta,0}(z-w)^{-2\alpha^2} + \text{regular terms}.$$

Clearly, the vertex operators are primary. In addition, they transform in a simple way under the action of the conserved current $J(z)$. Going back to the modes L_n and J_n, one can sharpen the notion of a primary scaling operator, by defining a scaling operator ϕ to be a $\widehat{u}(1)$ **Kac-Moody primary** if (i) $|\Delta\rangle = \phi(0)|0\rangle$ is an eigenstate of both L_0 and J_0 and (ii) $L_n|\Delta\rangle = J_n|\Delta\rangle = 0$ for all $n \geq 1$. The eigenvalue of J_0 is called a **charge**, hence the Kac-Moody primary operator V_α has the conformal weight $\Delta_{V_\alpha} = \alpha^2$ and charge α. Since one now has the further ladder operators J_{-n}, representations of Kac-Moody algebras combine several irreducible representations of the Virasoro algebra into a single irreducible representation.

One may readily extend the above results to find the following correlator

$$\langle J(z)V_{\alpha_1}(z_1)\ldots V_{\alpha_n}(z_n)\rangle = \left(\sum_{\ell=1}^{n} \frac{\alpha_\ell}{z-z_\ell} \right) \langle V_{\alpha_1}(z_1)\ldots V_{\alpha_n}(z_n)\rangle. \qquad (1.103)$$

From the Ward identities and the non-renormalisation theorems for conserved currents, one expects that for $|z| \gg 1$, one should have $J(z) \sim z^{-2}$. Combining this with the above correlator, one obtains the **neutrality condition** $\alpha_1 + \ldots + \alpha_n = 0$, since otherwise, $\langle V_{\alpha_1}(z_1)\ldots V_{\alpha_n}(z_n)\rangle$ must vanish. Applying the Wick theorem, the n-point correlator of the vertex operators reads

$$\langle V_{\alpha_1}(z_1)\ldots V_{\alpha_n}(z_n)\rangle = \prod_{i<j}(z_i - z_j)^{2\alpha_i\alpha_j} \delta_{\alpha_1+\ldots+\alpha_n,0}. \qquad (1.104)$$

The scaling operator $J\bar{J}$ plays a particular rôle, since its scaling dimension $x_{J\bar{J}} = 2$. It is therefore a **marginal scaling operator**. Theories with a marginal scaling operator contain a *line* of fixed points, along which the critical exponents may change as continuous functions of the scaling field conjugate to $J\bar{J}$. We shall encounter an example of this when discussing the $2D$ XY model below.

Proof In a heuristic way, the OPEs (1.102) can be derived as follows. From the Green's function of the free field $\varphi(z)$, one has by formal differentiation $\langle J(z)\varphi(w)\rangle = -i(z-w)^{-1}$. Then

$$J(z)V_\alpha(w) = \sum_{n=0}^{\infty} \frac{(i\alpha)^n}{n!} :J(z):\varphi(w)^n:$$

$$= \sum_{n=1}^{\infty} \frac{(i\alpha)^n}{n!}\frac{n}{i}\frac{1}{z-w}:\varphi(w)^{n-1}: + \text{regular terms}$$

$$= \frac{\alpha}{z-w}\sum_{n=1}^{\infty}\frac{(i\alpha)^{n-1}}{(n-1)!}:\varphi(w)^{n-1}: = \frac{\alpha}{z-w}V_\alpha(w).$$

The OPE with $T(z)$ is more subtle, but the repeated application of the OPE with $J(z)$ already shows that the most singular term is of the form $T(z)V_\alpha(w) \sim \frac{\alpha^2}{(z-w)^2}V_\alpha(w) + \ldots$ which permits to read off the conformal weight $\Delta_{V_\alpha} = \alpha^2$. The derivation of the other singular term in the OPE, which must be present, does require an explicit normal-ordering prescription for $:\varphi(w)^n:$ and is left to the reader. Finally, recall that for free, gaussian fields $\varphi_{1,2}$, one has the standard identity

$$:e^{a\varphi_1}::e^{b\varphi_2}: = :e^{a\varphi_1+b\varphi_2}:e^{ab\langle\varphi_1\varphi_2\rangle}.$$

Given the Green's function $\langle \varphi(z)\varphi(w)\rangle = -2\ln(z-w)$, a formal computation gives

$$V_\alpha(z)V_\beta(w) = :e^{i\alpha\varphi(z)+i\beta\varphi(w)}:e^{2\alpha\beta\ln(z-w)} = V_{\alpha+\beta}(w)(z-w)^{2\alpha\beta}$$

and where $\varphi(z) = \varphi(w) + (z-w)\partial\varphi(w) + \ldots$ was used, which beyond the first term merely produces further regular contributions. Application of the neutrality condition gives the last result announced in (1.102).   □

The Sugawara construction implies a characteristic relationship between the modes of the current and the energy-momentum tensor. These were defined as

$$J(z) = \sum_{n\in\mathbb{Z}} J_n z^{-n-1}, \qquad T(z) = \sum_{n\in\mathbb{Z}} L_n z^{-n-2}. \tag{1.105}$$

The normal-ordering prescription implies that two generators J_n and J_m commute under normal ordering, viz. $:J_n J_m: = :J_m J_n:$. Then a formal multiplication of the meromorphic series in $T(z) = :JJ:(z)$ leads to

$$L_n = \sum_{m\in\mathbb{Z}} :J_{n-m} J_n:. \tag{1.106}$$

Exercise 14 Given that $[J_n, J_m] = \frac{n}{2}\delta_{n+m,0}$, verify that the generators L_n in (1.106) satisfy the Virasoro algebra and re-derive the central charge $c = 1$ for the free boson.

The treatment of non-abelian Kac-Moody algebras $\widehat{g}_k$ requires a more involved and precise definition of the normal ordering than we have used here, since one can no more appeal to free-field-theory. For reference, we quote the result. If g is a semi-simple Lie algebra, one may from the structure constants define the metric tensor of g, according to $q_{ab} = f_{ad}{}^e f_{be}{}^d$. Since $\det q \neq 0$, the matrix q_{ab} has an inverse, denoted q^{ab}, such that $q_{ab}q^{bc} = \delta_a^c$. Then the structure constants $f_{abc} = q_{cd}f_{ab}{}^d$ are fully asymmetric $f_{abc} = -f_{bac} = -f_{acb}$. The '*dual Coexter number*' g is defined by $f^{abc}f_{dbc} = 2g\delta_d^a$. The non-abelian Sugawara form of the energy-momentum tensor is

$$T(z) = \frac{1}{2}\frac{1}{k+g}:q_{ab}J^a J^b:(z). \tag{1.107}$$

Then $J^a(z)$ is a primary operator with $\Delta_J = 1$. The central charge of the Virasoro algebra is

$$c = \frac{k \dim g}{k + g}. \tag{1.108}$$

The derivation of the relationship between the modes is left to the reader.

1.8 Compactifications and Modular Invariance

The action $S = \int d\bar{z} \wedge dz \, (\partial_{\bar{z}}\varphi)(\partial_z\varphi)$ of the free gaussian field $\varphi(z,\bar{z}) = \varphi(z) + \bar{\varphi}(\bar{z})$ is invariant under the transformations

$$\varphi \mapsto \varphi + \text{const.}, \quad \varphi \mapsto -\varphi \tag{1.109}$$

and one may use these symmetries to restrict the configuration space. For the simple model at hand, one may identify a **circle compactification** by identifying two configurations with fields φ and $\varphi + 2\pi\rho$, where ρ is the **compactification radius**. However, under a compactification the vertex operator correlations are unchanged if one takes $\alpha = n/\rho$, with $n \in \mathbb{Z}$.

A possible physical realisation is the $2D$ **XY model**, where on each site $\mathbf{i}$ of a square lattice is attached a continuous variable $\varphi_{\mathbf{i}} \in [0, 2\pi)$ and with the classical hamiltonian $\mathcal{H} = -J \sum_{(\mathbf{i},\mathbf{j})} \cos(\varphi_{\mathbf{i}} - \varphi_{\mathbf{j}})$, where the sum extends over pairs of nearest neighbours. Alternatively, one may introduce a two-component spin vector σ of unit length $|\sigma| = 1$ and the hamiltonian becomes $\mathcal{H} = -J \sum_{(\mathbf{i},\mathbf{j})} \sigma_{\mathbf{i}} \cdot \sigma_{\mathbf{j}}$. This model has an un-conventional second-order phase transition, not described by the usual power-laws but rather by essential singularities. The spin-spin correlator

$$G(\mathbf{r}) = \langle \sigma_{\mathbf{r}} \cdot \sigma_{\mathbf{0}} \rangle \sim \begin{cases} r^{-1/4} \ln^{1/8} r; & \text{if } T = T_c \\ r^{-\eta(T)}; & \text{if } T < T_c \end{cases} \tag{1.110}$$

remains algebraic in the entire low-temperature phase, with a temperature-dependent exponent $\eta(T)$, whereas $\lim_{T \to T_c} \eta(T) = \frac{1}{4}$. This comes about since the mechanism of this phase transition is not due to the standard formation of an ordered state but rather stems from the formation of *'vortices'*. Besides the vortex operators, there also exist so-called *'frustration lines'* whose end points are vortices. If the position **r** moves once around one of the end points of a frustration line, the value of the field $\varphi(\mathbf{r})$ increases by $2\pi\rho m$, with m being an integer. *Can one identify these operators in the framework of CFT?*

On the torus, which we have characterised by the complex **modular parameter** τ, the compactification condition is

$$\varphi\big(z + k + k'\tau\big) - \varphi(z) = 2\pi\rho\big(km + k'm'\big), \tag{1.111}$$

where the dependence on $\bar{z}$ is suppressed. One has the decomposition $\varphi = \varphi_{\text{per}} + \varphi_{\text{cl}}$ where φ_{per} is a fluctuating periodic field and (c.c. is the complex conjugate)

$$\varphi_{\text{cl}}(z) = 2\pi\rho\left(\frac{m\bar{\tau} - m'}{\bar{\tau} - \tau}z\right) + \text{c.c.} \tag{1.112}$$

is a solution of the equations of motion compatible with the compactification condition. The action is then $S = S_{\text{per}} + \pi\rho^2|m\tau - m'|^2/\operatorname{Im}\tau$. The model is thus decomposed into sectors described by the integers m, m'. The partition function of a sector is

$$Z_{m',m}(\rho,\tau) := Z_{\text{boson}}(\tau)\exp\left(-\pi\rho^2\frac{|m\tau - m'|^2}{\operatorname{Im}\tau}\right), \tag{1.113}$$

where the free boson partition function Z_{boson} was derived above in (1.84). Using the techniques explained in the previous section, one finds by a straightforward calculation that under a modular transformation

$$Z_{m',m}\left(\rho, \frac{a\tau + b}{c\tau + d}\right) = Z_{am'+bm,cm'+dm}(\rho,\tau). \tag{1.114}$$

Consequently, there are two obvious modular invariants. The first one is simply $Z_{0,0} = Z_{\text{boson}}$. The other one is obtained by summing over all values of m, m'

$$Z(\rho,\tau) := \rho\sum_{m',m} Z_{m',m}(\rho,\tau)$$

$$= (q\bar{q})^{-1/24}\frac{1}{P(q)P(\bar{q})}$$

$$\times \sum_{n,m}\exp\left[-2\pi i\operatorname{Re}\tau\cdot mn - 2\pi\operatorname{Im}\tau\frac{1}{2}\left(\rho^2 m^2 + \frac{n^2}{\rho^2}\right)\right], \tag{1.115}$$

after having applied the Poisson formula (1.82) to m'.[10] Since the partition function is a generating function for the primary and secondary states in the model, one sets $q = \exp(2\pi i\tau)$. Then, using (1.76), the scaling dimension $x_{n,m}$ and the spin $s_{n,m}$ of the primary operators $\mathcal{O}_{n,m}$ can be read off

$$x_{n,m} = \frac{1}{2}\left(\rho^2 m^2 + \frac{n^2}{\rho^2}\right), \qquad s_{n,m} = nm. \tag{1.116}$$

---

[10]Note that in this expression the factor $(\operatorname{Im}\tau)^{-1/2}$ of (1.84) has disappeared.

From these, one finds the conformal weights

$$\Delta_{n,m} = \frac{1}{4}\left(\frac{n}{\rho} + m\rho\right)^2, \qquad \overline{\Delta}_{n,m} = \frac{1}{4}\left(\frac{n}{\rho} - m\rho\right)^2 \qquad (1.117)$$

such that the partition function takes the final form

$$Z(\rho,\tau) = Z\left(\rho^{-1},\tau\right) = (q\bar{q})^{-1/24}\frac{1}{P(q)P(\bar{q})}\sum_{n,m=-\infty}^{\infty}q^{\Delta_{n,m}}\bar{q}^{\overline{\Delta}_{n,m}}. \qquad (1.118)$$

For $\rho > 1$, the quantum number n is said to describe *electric* fields, while the quantum number m is said to describe *magnetic* fields. Clearly, all conformal weights are ρ-dependent, but the spins are always integer.

The relationship with the XY model is as follows: the spin operator σ is identified with the conformal operator $\mathscr{O}_{\pm 1,0}$ such that the exactly known critical exponent $\eta = 2x_{\pm 1,0} = 1/4$ is reproduced. This implies the choice $\rho = 2$. The corresponding magnetic or vortex operator is then $\mathscr{O}_{0,\pm 1}$, implying $x_{0,\pm 1} = 2$. CFT therefore proposes a satisfactory way to explain the temperature-dependent exponent $\eta(T)$ is the low-temperature phase of the XY model, by a naturally realised marginal operator. If these identifications are valid, the central charge of the XY model is $c_{XY} = 1$.

1.9 From the Screened Coulomb Gas to Minimal Models

A modification of the free Coulomb gas, considered so far, is of interest. One changes the boundary conditions on the free field φ by introducing an additional vertex operator with charge $-2\alpha_0$ and conformal weight $\Delta_{-2\alpha_0}^{\text{free}} = 4\alpha_0^2$ placed at infinity. Then one considers the following modified correlators

$$\langle\!\langle V_{\alpha_1}(z_1)\dots V_{\alpha_n}(z_n)\rangle\!\rangle := \lim_{R\to\infty}R^{8\alpha_0^2}\langle V_{\alpha_1}(z_1)\dots V_{\alpha_n}(z_n)\,V_{-2\alpha_0}(R)\rangle. \qquad (1.119)$$

The limit is well-defined because the correlation function on the r.h.s. scales as $R^{-8\alpha_0^2}$ for large R. For these modified correlators, the neutrality condition becomes

$$\sum_{i=1}^{n}\alpha_i = 2\alpha_0. \qquad (1.120)$$

The non-vanishing two-point functions now read

$$\langle\!\langle V_\alpha(z)V_{2\alpha_0-\alpha}(0)\rangle\!\rangle = \frac{1}{z^{2\alpha(\alpha-2\alpha_0)}} \qquad (1.121)$$

and the conformal weight of the vertex operator V_α or $V_{2\alpha_0-\alpha}$ becomes

$$\Delta_\alpha = \Delta_{2\alpha_0-\alpha} = \alpha(\alpha - 2\alpha_0). \qquad (1.122)$$

It is non-trivial to show that this prescription is consistent, by establishing that the primary operators V_α and $V_{2\alpha_0-\alpha}$ are equivalent and have isomorphic Verma modules. The proof is beyond the scope of this book; we shall rather limit ourselves to list some interesting consequences and hope they sufficiently pique the reader's curiosity that he goes on and explores the rich literature on the subject for himself.

Due to the charge at infinity, the energy-momentum tensor must be modified, such that the vertex operator V_α remains a primary operator. The modified energy-momentum tensor is

$$T(z) = :JJ:(z) + 2\alpha_0 \partial_z J(z) = -\frac{1}{4}:\partial_z \varphi \partial_z \varphi:(z) + i\alpha_0 \partial_z^2 \varphi(z) \qquad (1.123)$$

with the same explicit normal-ordering prescription as before. The new term in the definition of T should not surprise because it is the only other possible conserved quantity with scaling dimension 2 available in our CFT. We remark that the same energy-momentum tensor also arises in Liouville field-theory.

The same techniques as used before can be used to show that both V_α and $V_{2\alpha_0 - \alpha}$ are primary operators (indeed, the constant in the extra term was chosen such that this is true). Furthermore, $T(z)$ satisfies the required OPE of an energy-momentum tensor, with the central charge

$$c = 1 - 24\alpha_0^2. \qquad (1.124)$$

This is the first remarkable result: the charge at infinity leads to a reduction of the central charge c. The second remarkable result is the existence of non-trivial **screening currents**, with conformal weight unity,

$$S_\pm(z) := V_{\alpha_\pm}(z) \quad \text{with } \alpha_\pm := \alpha_0 \pm \sqrt{\alpha_0^2 + 1}. \qquad (1.125)$$

These currents are used to create 'screening charges' which can screen the charge at infinity, defined as $Q_\pm := (2\pi i)^{-1} \oint_C dz\, S_\pm(z)$. They have a vanishing scaling dimension, but their application leads to a sift in the charge, by an amount of $\alpha_\pm$. Now, suppose that we restrict the subset of permissible vertex operators to the ones whose charge is quantised according to

$$\alpha_{r,s} = \frac{1}{2}(1 - r)\alpha_+ + \frac{1}{2}(1 - s)\alpha_- \qquad (1.126)$$

with $r, s \in \mathbb{N}$ are positive integers. Recall the parametrisation $c = 1 - 6/m(m+1)$ of the central charge. If we compute the conformal weight of $V_{r,s} := V_{\alpha_{r,s}}$ in terms of r, s, and m, one may recognise once more the Kac formula

$$\Delta_{\alpha_{r,s}} = \Delta_{r,s} = \frac{(r(m+1) - sm)^2 - 1}{4m(m+1)}. \qquad (1.127)$$

Therefore, *the operator content of the unitary minimal models is contained in the vertex operators of the screened Coulomb gas!* This correspondence can be further shown to lead to very useful integral representations for the n-point correlation functions, by balancing the total charge of the vertex operators by intelligently chosen screening operators. The details are beyond the scope of this introduction and we refer to the literature.

Exercise 15 Write down the modes L_n of the modified energy-momentum tensor in terms of the modes J_n of the conserved current. Find the central charge.

Fig. 1.7 Semi-infinite
geometry and complex
coordinates $z_a = u_a + iv_a$.
The physically inaccessible
lower half-plane is shaded
[26]

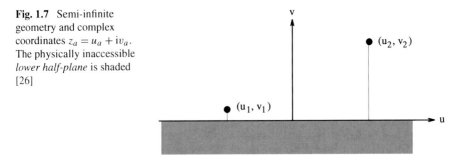

Exercise 16 Does the correspondence between the screened Coulomb gas and the primary operators $\phi_{r,s}$ of minimal models also apply to the non-unitary case, when $c = 1 - 6(p - p')/pp'$?

1.10 Surface Critical Phenomena

Boundary conformal field-theory is the study of conformal field-theory in domains with a boundary. Figure 1.7 illustrates the semi-infinite geometry which we shall study, together with the complex notation $z = u + iv$ to be used in $2D$. We begin with a short review on the thermodynamic behaviour near and at criticality and restrict to $2D$ from the outset.

In general, physical observables such as the position-dependent magnetisation profile $m = m(v)$, will depend on the distance v to the surface, besides the usual dependencies on the temperature T, an external magnetic field h and a surface magnetic field h_1 which only acts right at the boundary. A more complete notation would then be $m = m(v; T, h, h_1)$. One conventionally distinguishes the following quantities

$$m_b(T, h, h_1) = m(\infty; T, h, h_1) = -\frac{\partial g(T, h, h_1)}{\partial h} \qquad \text{bulk magnetisation}$$

$$m_s(T, h, h_1) = \int_0^\infty dv \left[m(v) - m_b \right] = -\frac{\partial g_s(T, h, h_1)}{\partial h} \qquad \text{excess magnetisation}$$

$$m_1(T, h, h_1) = m(0; T, h, h_1) = -\frac{\partial g_s(T, h, h_1)}{\partial h_1} \qquad \text{surface magnetisation}$$

where g_s is an excess contribution to the Gibbs potential density. Depending on the surface couplings and the surface field, there are in $2D$ two distinct possibilities: (i) the **ordinary transition**, when the surface magnetisation $m_1 < m_b$ and in particular, $m_1(T_c, 0, 0) = 0$ and (ii) the **normal transition**, when the surface magnetisation $m_1 \gtrsim m_b \neq 0$, even at criticality.[11] Turning to the description of the scaling in terms of critical exponents, one has (with $T \leq T_c$)

$$m_b(T, 0, 0) \sim (T_c - T)^\beta, \qquad m_s(T, 0, 0) \sim (T_c - T)^{\beta_s},$$
$$m_1(T, 0, 0) \sim (T_c - T)^{\beta_1} \tag{1.128}$$

where the exponent β of the bulk magnetisation is well-known and β_s, β_1 describe different aspects of scaling near to the surface. Scaling relations between these exponents and the standard bulk exponents are derived from the scaling form of the density of the excess Gibbs potential

$$g_s\left(\tau b^{y_\tau}, h b^{y_h}, h_1 b^{y_1}\right) = b^{-(d-1)} g_s(\tau, h, h_1) \tag{1.129}$$

where the exponent $d - 1$ reflects that the dimension of the surface is one less than in the bulk. This scaling form already shows that there is only a single new independent surface critical exponent. Indeed, it turns out that $\beta_s = \beta - \nu$ is related to bulk exponents, whereas β_1 cannot be reduced to bulk exponents alone. Bulk universality classes will therefore split into several surface universality classes which, among other elements, will be distinguished by the value of β_1.

Exercise 17 Define the excess specific heat and the exponent α_s by $C_s = -\partial^2 g_s/\partial \tau^2 \sim |\tau|^{-\alpha_s}$. Derive the scaling relation $\alpha_s = \alpha + \nu$ with the standard bulk critical exponents α, ν.

A last important observable is the two-point correlator of the order parameter σ

$$G(z_1, z_2) = \langle \sigma(u_1, v_1) \sigma(u_2, v_2) \rangle \sim \begin{cases} |z_1 - z_2|^{-\eta}; & \text{if } v_1, v_2 \to \infty \\ |u_1 - u_2|^{-\eta_\parallel}; & \text{if } v_1, v_2 \to 0 \end{cases} \tag{1.130}$$

which describes the cross-over between the two extreme cases (Fig. 1.7): bulk critical behaviour if the two points are far away from the surface and a new surface critical behaviour very close to the surface, with the new critical exponent $\eta_\parallel$. By analogy with $2D$ bulk criticality, where $\eta = 2x_\sigma = 2\beta/\nu$, one also introduces a **surface scaling dimension** $x_{\sigma,s}$ of the order parameter and then has

$$\eta_\parallel = 2x_{\sigma,s} = \frac{2\beta_1}{\nu}. \tag{1.131}$$

In contrast with the bulk, the surface scaling dimension of the energy density ε is simply $x_{\varepsilon,s} = 2$.

Example In contrast to ferromagnets, for *anti*-ferromagnetic systems the surface universality class can depend on the orientation of the surface with respect to the crystal axes. The $2D$ situation, to which we restrict here, is illustrated in Fig. 1.8.

---

[11] We do not require in this book the much more rich phenomenology of surface critical behaviour in $d \geq 3$ dimensions, with its 'extraordinary' and 'special' transitions.

Fig. 1.8 Surfaces *(10)* and
(11) for a square lattice, as
indicated by *dotted* and
dashed lines [26]

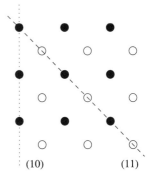

We shall illustrate this by considering an anti-ferromagnetic Ising model in external bulk and surface magnetic fields h and h_1, respectively. The classical hamiltonian is

$$\mathscr{H} = +J \sum_{\langle \mathbf{i},\mathbf{j} \rangle} \sigma_{\mathbf{i}} \sigma_{\mathbf{j}} - h \sum_{\mathbf{i}} \sigma_{\mathbf{i}} - h_1 \sum_{\mathbf{i},\mathrm{surf}}' \sigma_{\mathbf{i}}, \qquad (1.132)$$

where the last sum extends over the sites on the surface only. In the direction perpendicular to the surface, periodic boundary conditions are assumed. The ground
state of $\mathscr{H}$ is anti-ferromagnetic, as indicated by black and white sites in Fig. 1.8.
Hence both bulk and surface magnetic fields are non-ordering. The physical order
parameter is here the staggered magnetisation. Two cases must be distinguished:

(a) if the surface is along the (11) crystal direction, the symmetry of the model
under exchange of the 'black' and 'white' sub-lattices is kept intact. The problem
can then be mapped back to the usual ferromagnetic Ising model and one expects
ordinary surface critical behaviour, with $\eta_\| = 1$.

(b) On the other hand, if the surface is along the (10) direction, the sub-lattice
symmetry is broken. The order parameter profile is symmetric and antisymmetric
around the centre of a strip with L layers for L odd and even, respectively. If h
or h_1 are non-vanishing, this implies that the spins at the boundaries are fixed in a
relative $++$ or $+-$ orientation, for L odd or even, respectively. One thus expects a
normal surface transition, with $\eta_\| = 4$ in the $++$ case and $\eta_\| = 2$ in the $+-$ case.

These conclusion agree with the findings of BCFT, as we shall see in the next
section, as well as with numerical transfer matrix calculations [21].

1.11 Boundary Conformal Field-Theory

We begin with the projective conformal transformations in semi-infinite space. Of
the entire projective algebra $\mathfrak{sl}(2, \mathbb{R}) \oplus \mathfrak{sl}(2, \mathbb{R})$, only the so-called '*diagonal*' sub-

algebra isomorphic to $\mathfrak{sl}(2, \mathbb{R})$ which keeps the boundary line $v = 0$ invariant, are admissible. The generators read[12]

$$\ell_{-1} + \bar{\ell}_{-1} = -\partial_u$$
$$\ell_0 + \bar{\ell}_0 = -u\partial_u - v\partial_v - x \qquad (1.133)$$
$$\ell_1 + \bar{\ell}_1 = -(u^2 - v^2)\partial_u - 2uv\partial_v - 2xu.$$

We consider the two-point function $G(z_1, z_2) = \langle \phi_1(z_1)\phi_2(z_2) \rangle$ made from quasi-primary scaling operators $\phi_{1,2}(z)$ with scaling dimensions $x_{1,2}$ and the coordinates $z = u + iv$ as defined in Fig. 1.7. Projective conformal invariance predicts

$$G = \left(\frac{v_2}{v_1}\right)^{(x_2-x_1)/2} (v_1 v_2)^{-(x_1+x_2)/2} \Phi_{12}\left(\frac{(u_1 - u_2)^2 + (v_1 - v_2)^2}{v_1 v_2}\right). \qquad (1.134)$$

Here, the scaling function Φ_{12} remains an arbitrary differentiable function. In contrast to the conformally invariant two-point function in the bulk, there is *no* constraint on the scaling dimensions x_1 and x_2.

Two extreme cases can be recognised: first, if both points are far from the boundary, that is for $v_{1,2}$ both large, the two-point function should only depend on the distance $r^2 = (u_1 - u_2)^2 + (v_1 - v_2)^2$. In the limit $v_{1,2} \to \infty$, with r and v_2/v_1 both kept fixed, the explicit dependence on $v_1 v_2$ can be eliminated by assuming the behaviour $\Phi_{1,2}(U) \sim U^{-(x_1+x_2)/2} = U^{-\eta/2}$ for $U \to 0$. Finally, the explicit dependence on the orientation, via v_2/v_1, is eliminated by requiring that $x_1 = x_2$, if the bulk correlator is a non-vanishing function of r.[13] This reproduces the bulk result $G = g_0 \delta_{x_1,x_2} |\mathbf{r}|^{-\eta}$, with $\eta = 2x_\sigma = 2x_1$. On the other hand, if both points are close to the surface, one should rather look at the situation where $|u_1 - u_2| \to \infty$ and where now $v_{1,2}$ are kept fixed at finite values. Then the limit behaviour $\Phi_{12}(U) \sim U^{-\eta_\parallel/2}$ for $U \to \infty$ gives the correct phenomenology $G \sim |u_1 - u_2|^{-\eta_\parallel}$, and where the dependence on $v_{1,2}$ enters into the scaling amplitude. Projective conformal invariance alone cannot predict neither the form of the function $\Phi_{12}(U)$, nor fix the exponent $\eta_\parallel$.

Proof We outline the derivation of (1.134): from the projective conformal Ward identities

$$\sum_{i=1}^{2}\left[\frac{\partial}{\partial u_i}\right] G = 0$$

$$\sum_{i=1}^{2}\left[u_i \frac{\partial}{\partial u_i} + v_i \frac{\partial}{\partial v_i} + x_i\right] G = 0$$

$$\sum_{i=1}^{2}\left[(u_i^2 - v_i^2)\frac{\partial}{\partial u_i} + 2u_i v_i \frac{\partial}{\partial v_i} + 2x_i u_i\right] G = 0$$

---

[12]One restricts here and in what follows to scaling operators which are scalars deep in the bulk, with $\Delta = \bar{\Delta} = x/2$.

[13]It remains perfectly possible to describe by (1.134) a two-point function such as $\langle \sigma \varepsilon \rangle$ with a fixed non-vanishing magnetisation imposed at the surface and $x_\sigma \neq x_\varepsilon$, which of course vanishes in the bulk.

one has $G = G(u, v_1, v_2)$ with $u = u_1 - u_2$ and the two remaining conditions

$$u \frac{\partial G}{\partial u} + v_1 \frac{\partial G}{\partial v_1} + v_2 \frac{\partial G}{\partial v_2} + (x_1 + x_2)G = 0$$

$$\left(v_2^2 - v_1^2\right) \frac{\partial G}{\partial u} + u v_1 \frac{\partial G}{\partial v_1} - u v_2 \frac{\partial G}{\partial v_2} + u(x_1 - x_2)G = 0.$$

Now, change variables to $a = u^2/(v_1 v_2)$ and $b = v_1/v_2$ and let $G(u, v_1, v_2) = v_1^{-x_1} v_2^{-x_2} \Phi(a, b)$. Then the function $\Phi = \Phi(a, b)$ automatically satisfies the first of these, while the second one gives

$$\left[-\frac{\partial}{\partial a} + \frac{b^2}{b^2 - 1} \frac{\partial}{\partial b} \right] \Phi(a, b) = 0$$

and since a solution of this last equation is $a + b + b^{-1} - 2$, we arrive at (1.134). $\square$

An important question in a field-theory with boundaries is how to impose consistent boundary conditions. In $2D$ CFT, a natural requirement is that the off-diagonal component T_{uv} of the energy-momentum tensor should vanish, which intuitively means that *there is no momentum flow across the boundary*. For simplicity, one takes the domain to be the upper complex half-plane such that $T(z) = \bar{T}(\bar{z})$ when z is on the boundary, that is the real axis. Going again over the derivation of the conformal Ward identities Eqs. (1.32) and (1.35), one now has

$$\langle T(z)\phi_1(z_1, z_1') \ldots \phi_n(z_n, z_n') \rangle$$

$$= \sum_{j=1}^{n} \left(\frac{\Delta_j}{(z - z_j)^2} + \frac{1}{z - z_j} \frac{\partial}{\partial z_j} + \frac{\Delta_j}{(\bar{z} - z_j')^2} + \frac{1}{\bar{z} - z_j'} \frac{\partial}{\partial z_j'} \right)$$

$$\times \langle \phi_1(z_1, z_1') \ldots \phi_n(z_n, z_n') \rangle \qquad (1.135)$$

where the analytical continuation $T(z) = \bar{T}(\bar{z})$ for $\mathrm{Im}\, z < 0$ is used and where the points z_j' are the mirror images of the points z_j in the upper half-plane, reflected by the boundary, see Fig. 1.9, which also indicates the integration contours to be used. Consequently, the calculation of a two-point function in semi-infinite space reduces to the computation of a four-point function in the bulk.

Exercise 18 The conformal transformation $w = \frac{L}{\pi} \ln z$ maps the upper complex half-plane $\mathrm{Im}\, z \geq 0$ to an infinitely long strip of width L and with free boundary conditions.

(i) Show that the correlation length measured in the strip, at criticality, from the primary correlator $\langle \phi(w_1)\phi(w_2) \rangle \sim \exp[-\mathrm{Re}(w_1 - w_2)/\xi]$, has the finite-size scaling form $\xi = L/(\pi x_s)$, where x_s is the surface scaling dimension of the primary operator ϕ.

(ii) The density of the Gibbs potential on the strip has at criticality the form $g = g_0 + g_1 L^{-1} + g_2 L^{-2} + \ldots$. The bulk and surface contributions $g_{0,1}$ are non-universal. The universal amplitude g_2 is determined from the scaling of the excess Gibbs potential g_s. Show that $g_2 = -\pi c/24$.

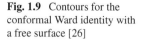

Fig. 1.9 Contours for the conformal Ward identity with a free surface [26]

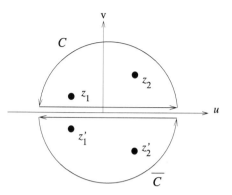

These results provide efficient algorithms for the calculation of x_s and c. Even better numerical efficiency can be achieved, however, if one uses a **Schwarz-Christoffel transformation** to conformally map the upper half-plane onto a rectangle with known aspect ratio and free boundary conditions [18].

Exercise 19 In semi-infinite space, **profiles** of scaling operators $\langle\phi(z)\rangle$ can be studied. If ϕ is quasi-primary, derive the profile $\langle\phi(v)\rangle = \phi_0 v^{-x}$, where x is the bulk scaling dimension of ϕ. If ϕ is primary, show further that in an infinitely long strip of finite width L and with free boundary conditions, the profile is

$$\langle\phi(v)\rangle = \phi_0\left[\left(\frac{L}{\pi}\right)\sin\left(\frac{\pi v}{L}\right)\right]^{-x},$$

where ϕ_0 is a normalisation constant.

Because of the analytic continuation to the lower half-plane applied to the energy-momentum tensor, there is only a single independent Virasoro algebra. Going again over the derivation of the quantum hamiltonian in (1.60), the conformal transformation $w = \frac{L}{\pi}\ln z$ from the upper half-plane to the open strip of width L leads to

$$H = \frac{1}{\pi}\int_0^L dv\, T(w) = \frac{\pi}{L}L_0 - \frac{\pi c}{24}\frac{1}{L}. \tag{1.136}$$

In what follows, a special feature of $2D$ boundary CFT will become important: it is possible to specify *different* boundary conditions on the positive and negative parts of the real axis, without modifying the essential structure of conformal field-theory. It can be shown that correlation functions behave as if an additional scaling operator, called '*boundary condition changing operator*' (**BCC operator**), had been inserted at the origin (even if this is not a conventionally local operator). In Fig. 1.10, the upper half-plane with two distinct boundary conditions, labelled a and b is indicated, and the dot symbolises the boundary condition changing operator. This geometry can be conformally mapped onto an annulus, which may also be formed from a rectangle of width L and of height δ, and with the top and bottom edges identified, see Fig. 1.10. In the case of infinite height, that conformal transformation will be the

Fig. 1.10 Conformal
mapping from the
semi-infinite space, with two
distinct boundary conditions
labelled a and b, to the
annulus

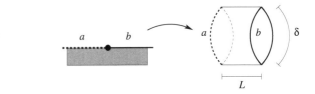

usual one, viz. $w = \frac{1}{\pi} \ln z$, such that the quantum hamiltonian $H = H_{ab}$ which may
also be viewed as the generator of translations along the strip, is given by (1.136).

Similarly, for the annulus of unit width $L = 1$, where one sets $q = e^{-\pi \delta}$, one has
the partition function

$$Z_{ab}(\delta) = \operatorname{tr} e^{-\delta H_{ab}} = \operatorname{tr} q^{L_0 - \pi c/24} = \sum_\Delta n_{ab}(\Delta) \chi(\Delta, c) \qquad (1.137)$$

which, analogously to the periodic case treated before, can be decomposed into
a sum of characters $\chi(\Delta, c)$. Here, the positive integers $n_{ab}(\Delta)$ give the operator
contents with the boundary conditions ab, such that the lowest admissible value of
Δ such that $n_{ab}(\Delta) > 0$ gives the conformal weight of the BCC operator. The other
admissible values of Δ give the conformal weight of those primary operators which
make up the partition function.

On the other hand, the annulus partition function, up to an overall rescaling, can
be interpreted as the path integral of a conformal field-theory on a circle of unit
circumference, for an imaginary time $1/\delta$. Then the partition function becomes a
matrix elements between the **boundary states** $\langle a|$ and $|b\rangle$:

$$Z_{ab}(\delta) = \langle a|e^{-H/\delta}|b\rangle \qquad (1.138)$$

but where now the quantum hamiltonian is given by (1.60) and the boundary states
are taken from the Verma module $\mathscr{V}$. *What are the allowed boundary states and
which are the values of $n_{ab}(\Delta)$?*

We only present but the most brief of an outline and refer to later chapters in this
volume and the literature for details. A first condition arise from the requirement of
a vanishing momentum flux across the boundary, $T(z) = \bar{T}(\bar{z})$ for z real. Inserting
the mode expansion and transforming onto the cylinder geometry, it can be shown
that for any boundary state $|B\rangle$ one must have

$$L_n|B\rangle = \bar{L}_{-n}|B\rangle \qquad (1.139)$$

and furthermore, because of the decomposition of the Verman module $\mathscr{V}$, this con-
dition must hold in each subspace $\mathscr{V}_{\Delta_B} \oplus \mathscr{V}_{\bar{\Delta}_B}$. For example, taking $n = 0$, means
that boundary states must be scalar, $\Delta_B = \bar{\Delta}_B$.

It turns out that these constraints on the admissible boundary states (called
Ishibashi states) can be fully solved, and in a beautiful way. This holds true at
least for the so-called diagonal models, where in the Verma module decomposition
Eq. (1.73) $n_{\Delta, \bar{\Delta}} = \delta_{\Delta, \bar{\Delta}}$, but also for all non-diagonal minimal models. Several important technical assumptions have to be made at this point, which we skip. We
introduce a slightly modified notation. The boundary condition a, which we have

already seen to be characterised by a scalar boundary state, will be referred to by their conformal weight $\Delta_a = \overline{\Delta}_a$. Then, we re-write the multiplicity in the partition function Eq. (1.137) as $n_{ab}(\Delta) = n_{\Delta_a, \Delta_b}^{\Delta}$. Recall from the section on modular invariance (see p. 23) the fusion algebras, with the fusion coefficients re-written as $N_{ab}^c = N_{\Delta_a, \Delta_b}^{\Delta_c}$. Now, **Cardy's formula** gives the operator content of $2D$ boundary conformal field-theories:

$$n_{\Delta_a, \Delta_b}^{\Delta} = N_{\Delta_a, \Delta_b}^{\Delta}. \tag{1.140}$$

Example We illustrate this profound relation in the $2D$ Ising universality class. Before, the identifications $\mathbf{1} = \chi_{1,1} = (0)$, $\sigma = \chi_{1,2} = (\frac{1}{16})$ and $\varepsilon = \chi_{2,1} = (\frac{1}{2})$ have been established, where we now suppress the second conformal weight, since all the primary operators are scalars. We shall need the following fusion rules

$$\sigma \odot \sigma = \mathbf{1} + \varepsilon, \qquad \mathbf{1} \odot \mathbf{1} = \varepsilon \odot \varepsilon = \mathbf{1},$$
$$\varepsilon \odot \mathbf{1} = \varepsilon, \qquad \sigma \odot \mathbf{1} = \sigma \odot \varepsilon = \sigma. \tag{1.141}$$

In the $2D$ Ising model, one may define the following boundary conditions: (i) free (F), where the order parameter vanishes at the boundary, (ii) fixed ($+$ or $-$), where the local order parameter is constrained to $\sigma|_{\text{boundary}} = \pm 1$. On an infinitely long strip of finite width L, one then has the following combinations:

(a) free boundary conditions on both sides. Then the partition function is denoted by $Z^{(F)} = Z^{(FF)}$;

(b) fixed boundary conditions, with the spins fixed either in the same way or in the opposite way, which gives the partition functions $Z^{(++)} = Z^{(--)}$ and $Z^{(+-)} = Z^{(-+)}$;

(c) mixed boundary conditions, which are free on one side and fixed on the other, with the partition functions $Z^{(M)} := Z^{(F+)} = Z^{(F-)} = Z^{(+F)} = Z^{(-F)}$.

The correspondence between the primary operators and the boundary conditions is listed in the following table:

primary	boundary condition
σ	free (F)
$\mathbf{1}$	fixed ($+$)
ε	fixed ($-$)

but the correspondence between the operators $\mathbf{1}, \varepsilon$ and the boundary conditions $(+), (-)$ can also be permuted.

With this correspondence, and the explicit fusion rules (1.141), Cardy's formula then yields the following predictions for the operator content:

$$Z^{(F)} = \chi_{1,1} + \chi_{2,1} = (0) + \left(\frac{1}{2}\right) \qquad \text{ordinary transition, } \eta_{\parallel} = 1$$

$$Z^{(++)} = Z^{(--)} = \chi_{1,1} = (0) \qquad\qquad \text{normal transition, } \eta_{\parallel} = 4$$

$$Z^{(+-)} = Z^{(-+)} = \chi_{2,1} = \left(\frac{1}{2}\right) \qquad \text{normal transition, } \eta_{\parallel} = 2$$

$$Z^{(M)} = \chi_{1,2} = \left(\frac{1}{16}\right) \qquad\qquad \text{mixed transition, } \eta_{\parallel} = 2$$

$$(1.142)$$

in full agreement either with an exact solution of the $2D$ Ising model or else numerical studies. We also indicate the kind of the surface transition and list the resulting values of $\eta_{\parallel} = 2x_{\sigma,s}$, where the surface exponent can be read off from the smallest gap $E_1 - E_0 = \frac{\pi}{L}x_{\sigma,s}$ of the quantum hamiltonian H_{ab} on the strip. Alternatively, $x_{\sigma,s}$ can be read off from the Rocha-Caridi formulæ.

Example Infinitely long defect line(s) in the $2D$ Ising model play a special rôle, since such perturbations are *marginal* and the critical exponents near to the defect line(s) depend continuously on the defect strengths κ_i. Equivalent situations are found when several sectors of Ising models are coupled at their respective boundaries. This goes beyond the situation described so far. While the semi-infinite Ising model is a minimal model with a finite number of primary operators, the Ising model with defect lines contains an *infinite* number of primary states! It can be shown that the partition function, in the case of a single defect line, can be expressed in terms of the characters of a 'shifted' $\mathfrak{u}(1)$-Kac-Moody algebra [3, 27], which also takes into the account the continuous shift of the scaling dimensions with the defect strengths. We merely mention that this situation can also be included into the framework of boundary conformal field-theory, with relations to so-called '*orbifold compactifications*' and related to the Ashkin-Teller model [33]. Physically, such defects lines can conviently realised when studying the corner magnetisation at the surface transition in $3D$ Ising models with enhanced surface couplings [36] or at the border line of two distinct Ising models, even with different spins [31].

1.12 Notes and References

Two-dimensional conformal invariance started with the breakthrough article by Belavin, Polyakov and Zamolodchikov [5]. Since then, many excellent reviews and books on conformal invariance have been published. Presently, the most comprehensive of them certainly is the superb book written by di Francesco, Matthieu and Sénéchal [19].

We have assumed that the reader is familiar with the basic notions of phase transitions and critical phenomena, at thermal equilibrium. Taking off almost from the very beginning, the nicely written book [40] is an excellent starting point for the novice. A masterly and lucid introduction to the main ideas of scaling, universality and the renormalisation-group, and much more, can be found in a set of superb lectures [23]. A magnificent modern summary of the scaling description of critical phenomena, as they occur in a huge variety of different settings, and leading up to conformal invariance, is given in [15]. See [29] for a brief introduction to critical phenomena with particular attention to the non-universal metric factors, along with long lists of critical exponents. In [26], the reader can find, besides an introduction to the elements of conformal invariance, many experimental and numerical illustrations, including long lists of measured critical exponents, of $2D$ critical phenomena and various explicit tests of many aspects of $2D$ conformal invariance. For reviews on surface critical phenomena, consult [20, 35]. Those interested in integrable systems will naturally start consulting [4], or [24].

Much of what we know about conformal invariance, we have learned on one hand from the many important and often so crystal-clear articles and reviews written by Cardy [11, 14, 17] and on the other hand from the beautiful lectures which achieved their final form in [22, 30]. The lectures written up in [41] were a useful source of inspiration. Readers interested in the mathematical aspects of conformal invariance may consult [39]. A nice modern introduction, slightly oriented towards string theory, is [6]. Some of the most simple elements of conformal invariance are presented in a very accessible manner in [25], for *francophone* readers. A nice classical summary of non-renormalisation theorems in quantum field-theory is [7].

Boundary conformal field-theory arose largely in [10, 12]. The relationship to fusion algebras and the Verlinde formula appears in [13]. For detailed reviews on this, see [16] and [34]. The application of boundary conformal field theory to antiferromagnetic Ising models is taken from [21].

For a reader curious about whether at least some ideas of conformal invariance might have a fruitful bearing on dynamics and non-equilibrium phase transitions, we suggest to consult [28], which looks at the dynamics of *classical* quenched systems. *Quantum* quenches and their relationship with conformal invariance are introduced in [8, 9].

References

1. Alcaraz, F.C., Grimm, U., Rittenberg, V.: The XXZ Heisenberg chain, conformal invariance and the operator content of $c < 1$ systems. Nucl. Phys. B **316**, 735 (1989)
2. Alcaraz, F.C., Levine, E., Rittenberg, V.: Conformal invariance and its breaking in a stochastic model of a fluctuating interface. J. Stat. Mech., 08003 (2006)
3. Baake, M., Christe, P., Rittenberg, V.: Higher spin conserved currents in $c = 1$ conformally invariant systems. Nucl. Phys. B **300**, 637 (1988)
4. Baxter, R.J.: Exactly Solved Models in Statistical Mechanics. Academic Press, London (1982)
5. Belavin, A.A., Polyakov, A.M., Zamolodchikov, A.B.: Infinite conformal symmetry in two-dimensional quantum field-theory. Nucl. Phys. B **241**, 333 (1984)
6. Blumenhagen, R., Plauschinn, E.: Introduction to Conformal Field-Theory. Lecture Notes in Physics, vol. 779. Springer, Heidelberg (2009)
7. Boyer, T.H.: Conserved currents, renormalization and the Ward identity. Ann. of Phys. **44**, 1 (1967)
8. Calabrese, P., Cardy, J.L.: Time-dependence of correlation functions following a quantum quench. Phys. Rev. Lett. **96**, 136801 (2006)

9. Calabrese, P., Cardy, J.L.: Entanglement and correlation functions following a local quench: a conformal field theory approach. J. Stat. Mech., 10004 (2007)
10. Cardy, J.L.: Conformal invariance and surface critical behaviour. Nucl. Phys. B **240**, 514 (1984)
11. Cardy, J.L.: Conformal invariance. In: Domb, C., Lebowitz, J.L. (eds.) Phase Transitions and Critical Phenomena, vol. 11. Academic Press, London (1986)
12. Cardy, J.L.: Effect of boundary conditions on the operator content of two-dimensional conformally invariant theories. Nucl. Phys. B **275**, 200 (1986)
13. Cardy, J.L.: Boundary conditions, fusion rules and the Verlinde formula. Nucl. Phys. B **324**, 581 (1989)
14. Cardy, J.L.: Conformal invariance and statistical mechanics. In: Brézin, E., Zinn-Justin, J. (eds.) Fields, Strings and Critical Phenomena, Les Houches XLIX. North-Holland, Amsterdam (1990)
15. Cardy, J.L.: Scaling and Renormalization in Statistical Physics. Cambridge University Press, Cambridge (1996)
16. Cardy, J.L.: Boundary conformal field theory. In: Encyclopedia of Mathematical Physics. Elsevier, Amsterdam (2006)
17. Cardy, J.L.: Conformal field theory and statistical mechanics. In: Exact Methods in Low-Dimensional Statistical Physics and Quantum Computing, Les Houches XLIX. North-Holland, Amsterdam (2008)
18. Chatelain, C., Berche, B.: Tests of conformal invariance in randomness-induced second-order phase transitions. Phys. Rev. E **58**, 6899 (1998)
19. di Francesco, P., Mathieu, P., Sénéchal, D.: Conformal Field-Theory. Springer, Heidelberg (1997)
20. Diehl, H.W.: Field-theoretical approach to critical phenomena at surfaces. In: Domb, C., Lebowitz, J.L. (eds.) Phase Transitions and Critical Phenomena, vol. 10. Academic Press, London (1987)
21. Drewitz, A., Leidl, R., Burkhardt, T.W., Diehl, H.W.: Surface critical behaviour of binary alloys and antiferromagnets: dependence of the universality class on surface orientation. Phys. Rev. Lett. **78**, 1090 (1997)
22. Drouffe, J.-M., Itzykson, C.: Statistical Field-Theory, vol. 2. Cambridge University Press, Cambridge (1988)
23. Fisher, M.E.: Scaling, universality and renormalisation group theory. In: Hahne, F.J.W. (ed.) Critical Phenomena. Lecture Notes in Physics, vol. 186, pp. 1–13. Springer, Heidelberg (1983)
24. Gaudin, M.: La Fonction d'onde de Bethe. Masson, Paris (1983)
25. Grandati, Y.: Éléments d'introduction à l'invariance conforme. Ann. Physique **17**, 159 (1992)
26. Henkel, M.: Phase Transitions and Conformal Invariance. Springer, Heidelberg (1999)
27. Henkel, M., Patkós, A.: Critical exponents of defective Ising models and the $U(1)$ Kac-Moody-Virasoro algebras. Nucl. Phys. B **285**, 29 (1987)
28. Henkel, M., Pleimling, M.: Non-equilibrium Phase Transitions, vol. 2. Ageing and Dynamical Scaling Far from Equilibrium. Springer, Heidelberg (2010)
29. Henkel, M., Hinrichsen, H., Lübeck, S.: Non-equilibrium Phase Transitions, vol. 1. Absorbing Phase Transitions. Springer, Heidelberg (2009)
30. Itzykson, C., Drouffe, J.-M.: Théorie Statistique des Champs, vol. 2. InterÉditions/CNRS, Paris (1989)
31. Karevski, D., Henkel, M.: Finite-size effects in layered magnetic systems. Phys. Rev. B **55**, 6429 (1995)
32. Neto, A.H.C., Guinea, F., Peres, N.M.R., Novosolev, K.S., Geim, A.K.: The electronic properties of graphene. Rev. Mod. Phys. **81**, 109 (2009)
33. Oshikawa, M., Affleck, I.: Boundary conformal field theory approach to the two-dimensional critical Ising model with a defect line. Nucl. Phys. B **495**, 533 (1997)
34. Petkova, V., Zuber, J.-B.: Conformal boundary conditions and what they teach us. In: Horváth, Z., Palla, L. (eds.) Non-perturbative Quantum Field Theoretic Methods and Their Applications. World Scientific, Singapore (2001)

35. Pleimling, M.: Critical phenomena at perfect and non-perfect surfaces. J. Phys. A, Math. Gen. **37**, 79 (2004)
36. Pleimling, M., Selke, W.: Ising cubes with enhanced surface couplings. Phys. Rev. E **61**, 933 (2000)
37. Polchinski, J.: Scale and conformal invariance in quantum field theory. Nucl. Phys. B **303**, 226 (1988)
38. Riva, V., Cardy, J.L.: Scale and conformal invariance in field theory: a physical counterexample. Phys. Lett. B **622**, 339 (2005)
39. Schottenloher, M.: A Mathematical Introduction to Conformal Field-Theory. Lecture Notes in Physics, vol. 759. Springer, Heidelberg (2008)
40. Yeomans, J.M.: Statistical Mechanics of Phase Transitions. Oxford University Press, Oxford (1992)
41. Zuber, J.-B.: An introduction of conformal field-theory. Acta Phys. Pol. B **26**, 1785 (1995)

Chapter 2
A Short Introduction to Critical Interfaces in 2*D*

Michel Bauer

2.1 Introduction

The central goal of this chapter is to introduce stochastic Loewner evolution (SLE), but with a detailed emphasis on its interplay with statistical mechanics and conformal field-theory.

Stochastic Loewner evolutions describe growth processes, and as such they fall in the more general category of **growth phenomena**. These are ubiquitous in the real world at all scales, from crystals and plants to dunes and galaxies, and so on. They can be addressed in many ways, by deterministic or probabilistic methods, in discrete or continuous space and time. Understanding growth is usually a very difficult task. This is true even in two dimensions, the only case we mention in these notes. Yet in two dimensions, the powerful tools of complex analysis allow to tame the zoo of shapes. Indeed, many relevant growth processes involve the growth of domains (i.e. contractile open subsets of the Riemann sphere). *Riemann's uniformisation theorem* describes domains in a canonical way by conformal maps, and then growth of domains by Loewner chains. So at least the kinematic part is "easy" in two dimensions. This, and more, is explained in details in Sect. 2.3. To avoid any confusion, let us stress that being able to describe a growth process using tools from complex analysis and conformal geometry does not mean that the growth process itself is conformally invariant at all.

The mathematics in Sect. 2.2 are more down to earth. They involve mostly simple combinatorics. We concentrate on two simple examples of random geometric curves on the lattice, the *exploration process* and the *loop-erased random walk*. Both examples have been shown to have a continuum limit, moreover described by SLE.

M. Bauer (✉)
IPhT, CEA-Saclay, 91191 Gif-sur-Yvette, France
e-mail: michel.bauer@cea.fr

M. Bauer
LPTENS, 24 rue Lhomond, 75005 Paris, France

M. Henkel, D. Karevski (eds.), *Conformal Invariance: an Introduction to Loops, Interfaces and Stochastic Loewner Evolution*, Lecture Notes in Physics 853, DOI 10.1007/978-3-642-27934-8_2, © Springer-Verlag Berlin Heidelberg 2012

We explain how to make numerical simulations, and in particular how to estimate fractal dimensions. We recall how both the exploration process and the loop-erased random walk have a natural interpretation as interfaces in some (critical) statistical mechanics models called *loop models*. The exploration process leads easily to *percolation*, while loop-erased random walks lead to a more intricate combinatorics related to determinantal identities and symplectic fermions.

We also mention in passing another important but still very mysterious random growth model, *diffusion-limited aggregation* (DLA), which is expected to have a non-conformally invariant continuum limit. Simulations of DLA produce beautiful shapes, as do many other growth models. The universality classes are still debated, but DLA and its cousins, *Laplacian Growth*, the *Hele-Shaw problem*, *dielectric breakdown* are all important in various applications, though we shall not touch this subject in these notes.

Stochastic Loewner evolution is derived in Sect. 2.4, following Schramm's original argument. It's validity rests on two properties: *conformal invariance* which can be expected only in the continuum limit, and the *domain Markov property*, which holds already on the lattice in many models (including the exploration process and loop-erased random walks). The outcome if that conformally invariant measures on curves with the domain Markov property are parameterised by a single arbitrary positive number, named κ, which appears as the variance parameter of a Brownian motion.

SLE is a simple but particularly interesting example of growth process for which the growth is local and continuous so that the resulting set is a continuous curve, or at least is closely related to a continuous curve. What makes stochastic Loewner evolutions so important (at least in the author's view) is first that they are among the very few growth processes that can be studied analytically in great detail, and second that they have solved a problem that remained opened for two decades, the description of conformally invariant extended objects. For a physicist working in statistical mechanics, in particular for a conformal field-theorist, a rather startling feature of SLE is that is shows the *hidden Markovian character of critical interfaces*, which is far from obvious *a priori*. At the technical level, another striking feature of SLE is that in has turned many questions concerning interfaces that seemed just out of reach into exercises in stochastic calculus. Of course this simplicity is one of the blueprints of important discoveries. To give one illustration, we show how the locality property of the exploration process leaves $\kappa = 6$ as the sole possibility to describe its continuum limit.

The last section, Sect. 2.5 is devoted to the basics of the relationships between SLE and conformal field-theory (CFT). It is reassuring that the two subjects are closely interwoven. The basic statement is that conformal field-theory and stochastic Loewner evolutions are coupled in such away that CFT correlators are SLE martingales. The origin of this relationship is basically an instance of double counting, as we show in the last subsection. The consequence is that the Itô generator of SLE seen as a diffusion has to coincide with a particular singular vector differential operator. This allows to retrieve easily the relation between the central charge c (a CFT characteristic) and the SLE parameter κ:

$$2\kappa c = (6 - \kappa)(3\kappa - 8). \tag{2.1}$$

A more general consequence is that SLE probabilities have all the axiomatic properties of CFT correlators involving a special boundary-changing operator of dimension $\Delta = (6 - \kappa)/2\kappa$ inserted at the origin of the interface. We illustrate this by computing the simplest hitting probability either via stochastic calculus or via operator product methods. We also emphasise the importance of CFT partition functions to describe variants of SLE involving conditioning for instance, and more generally as a guiding principle to study SLE. Of course, physicists understand the relevance of partition functions because of their training in statistical mechanics, but this view is also shared by mathematicians now.

In the last paragraph, we wrote "*SLE probabilities have all the axiomatic properties of CFT correlators*" which may seem quite a twisted statement. Writing simply "SLE probabilities are CFT correlators" is OK with a nasty drawback: the CFTs of SLEs are badly behaved ones. Thirty years of CFT have make physicists comfortable with rational CFT, or unitary CFTs. The CFTs of SLEs are yet to be precisely defined, but they are definitely neither rational nor unitary. They are probably closest to certain logarithmic CFT. The operator content, the fusion algebra, etc, are unknown at the moment. The only correlators that are under control via general conformal invariance arguments are those with at most 4 boundary operators (or 1 bulk and 2 boundary operators). This may seem very negative. However, SLE is a perfectly well-defined mathematical object, and one can hope that this will give enough control to learn some general lessons on what CFT is about, outside the reassuring but limited regions of unitarity and/or minimality. There has been some recent progress in this direction.

2.2 Discrete Models

Random curves have focused the interest of physicists and mathematicians for decades. The simplest and perhaps oldest example is the symmetric **random walk** on the lattice or its continuous counterpart, **Brownian motion**. For dimensions ≤ 4 it is not a simple curve. On the other hand, polymers have a strong tendency to be *self-avoiding*, and they can be modelled crudely as simple random walks with a statistical weight giving fugacity μ to each monomer. But there is a wealth of interesting models of simple random walks. Among them are interfaces in $2D$ systems. Under certain circumstances, such systems are expected to have a continuum limit.

Recently a lot of progress has been achieved. Notably, *a classification of random curves in the continuum with certain special properties has been obtained*. It has received the name "**Stochastic Loewner Evolutions**" (SLE), and it is the subject of Sect. 2.4. It is hard to over-estimate the impact of SLE: it has given tools to solve formidable problems by routine computations, but moreover it has made it possible to prove that families of random simple walks and interfaces have a continuum limit.

The purpose of the examples that follow is to illustrate the connection between geometrical random curves and statistical mechanics. It turns out that partition functions under various disguises play a huge role in the study of SLE.

Fig. 2.1 *Left panel*: a smooth domain. *Middle panel*: a non-simply connected region is not a domain. *Right panel*: a non-smooth domain, whose topological boundary is not a simple curve

2.2.1 Discrete Domains

In what follows, a **domain** $\mathbb{D}$ is a non-empty open simply connected (i.e. no holes) strict subset of the complex plane $\mathbb{C}$. With this generality, domains and their topological boundaries can be quite complicated, as exemplified by the domain on the right in Fig. 2.1. There are mathematical theories to define a better notion of boundary suitable for our purposes. This can be achieved via the theory of "prime ends" or via the so-called *Poisson/Martin boundaries* that parameterise harmonic functions i.e. solutions h of the Laplace equation $\Delta_{\mathbf{r}} h = 0$, but we shall say only a few words about that in Sect. 2.3.1.

The complex plane admits regular tilings by hexagons, by triangles or by squares. The following definitions are given for hexagonal tilings, but they can easily be adapted for tilings by triangles and squares.

All hexagonal **tiling**s can be obtained from one of them by similarities (in complex notation $z \mapsto \lambda z + \rho$). Fix such a tiling $\mathcal{T}$, for instance one whose hexagons have unit area. The plane is the disjoint union of vertices, open edges and open faces of $\mathcal{T}$: every point in the plane is either a **vertex**, or an **interior point** of an edge, or an interior point of a face.

A **hexagonal domain** $\mathbb{D}$ with reference $\mathcal{T}$ is a domain in the usual sense as defined above which is the union of vertices, open edges and open faces of $\mathcal{T}$.

This definition accommodates "smooth" domains like the left one in Fig. 2.2 whose boundary is a simple curve but also more irregular shapes like the middle one in Fig. 2.2 whose boundary is not a simple curve. If $\varepsilon > 0$ is much smaller than the size of an edge of $\mathcal{T}$, the points in the hexagonal domain $\mathbb{D}$ whose distance to the complement of $\mathbb{D}$ is ε form a simple curve, but the limit $\varepsilon \to 0^+$ is singular. The knowledge of the side from which a boundary point is approached is naively lost in the limit, but one can decide to keep track of it and this is the most useful definition of boundary in this context. For hexagonal domains we have thus a notion of **boundary** which makes it a curve even for a non-smooth domain. That such a boundary can also be defined for general domains is a non-trivial matter.

In these notes, an **admissible boundary condition** is a couple of distinct vertices (a, b) of $\mathcal{T}$, $a, b \notin \mathbb{D}$ such that there is (at least) a path from a to b in $\mathbb{D}$. A **path** (or **simple walk**) in $\mathbb{D}$ is a sequence $s_1, \ldots, s_{2n+1}$, where $a = s_1$, $b = s_{2n+1}$, the s_{2m+1}, $1 \leq m < n$, (if any) are distinct vertices of $\mathcal{T}$ in $\mathbb{D}$ and the s_{2m}, $1 \leq m < n$, are

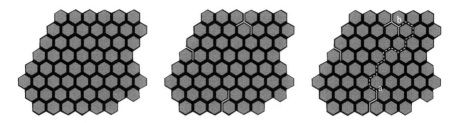

Fig. 2.2 *Left panel*: a "smooth" hexagonal domain. *Middle panel*: a non-smooth hexagonal domain. *Right panel*: an admissible boundary condition

distinct edges of $\mathscr{T}$ in $\mathbb{D}$ with boundary $\{s_{2m-1}, s_{2m+1}\}$. This is illustrated on the left of Fig. 2.2. Any such path splits $\mathbb{D}$ into a left and a right piece. The term "simple walk" is also used to mean a "path".

If $s_1, \ldots, s_{2n+1}$ is a path from a to b in $\mathbb{D}$ and $0 \le m < n$, the set $\mathbb{D}'$ obtained by removing from $\mathbb{D}$ the sets s_l, $1 < l \le s_{2m+1}$ is still a domain, and (s_{2m+1}, b) is an *admissible boundary condition* for $\mathbb{D}'$.

Our main interest in the next subsections will be in measures on paths from a to b in $\mathbb{D}$ when $\mathbb{D}$ is a domain and (a, b) an admissible boundary condition.

Hexagonal domains have a *special property* which is crucial for what follows. Suppose $(\mathbb{D}, a, b)$ is a hexagonal domain with admissible boundary condition. The right (resp. left) hexagons are by definition those which are on the right (resp. left) of every path from a to b in $\mathbb{D}$. Left and right hexagons are called **boundary hexagon**s. The other hexagons of $\mathbb{D}$ are called **inner hexagons**.[1] Colour the left hexagons in black (say) and the right hexagons in white as in Fig. 2.3 on the left. If one colours the inner hexagons arbitrarily in black or white, then *there is a single path from a to b in* $\mathbb{D}$ *such that the hexagon on the left (resp. right) of any of its edges is black (resp. white)*. This is illustrated in Fig. 2.3 on the right. This path can be defined recursively because a is on the boundary of at least one left and at least one right hexagon: as a is not in $\mathbb{D}$, in any colouring there is exactly one edge in $\mathbb{D}$ with a on its boundary and bounding two hexagons of different colours. Start the path with this edge and go on.

If $\mathbb{D}$ is domain with a smooth boundary, it is easy to approximate it with high precision by hexagonal domains with reference tiling $\lambda \mathscr{T} + \rho$ by taking λ small enough. A general domain $\mathbb{D}$ may have a very complicated boundary, and approximations by hexagonal domains is not so obvious. Despite their importance, we shall remain silent on these subtleties. Also, it is useful for the general theory to have a quantitative notion of how close such an approximation is to the original domain and to have quantitative notion of *convergence* of approximations (when λ gets smaller and smaller) that guaranties that some phenomena on discrete domains (for instance some properties of certain statistical mechanics models) have an interpretation in the limit. We shall not give a formal definition of convergence, but simply mention that it exists.

---

[1]Note that being a boundary or an inner hexagon depends on (a, b).

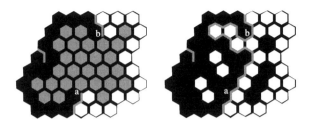

Fig. 2.3 *Left panel*: the boundary of a hexagonal lattice domain with boundary conditions. *Right panel*: the interface associated to a configuration

All the examples of **interface**s we shall deal with in these notes can be defined on arbitrary hexagonal domains with admissible boundary condition, though sometimes we shall use square domains. Certain geometrical examples will define directly a law for the interface or a probabilistic algorithm to construct samples. Examples from statistical mechanics will give a weight for each colouring of the inner hexagons, and the law for the interface can be derived (at least in principle) from this more fundamental weight. The model of interface can depend on some parameters, called collectively p (for instance, temperature can be one of those).

Consider an interface model with parameter family p on discrete domains. Fix a domain with two marked boundary points, $(\mathbb{D}, a, b)$ and suppose it can be approximated by a "convergent" sequence of discrete domains with boundary conditions $(\mathbb{D}_n, a_n, b_n)$ whose reference tiling $\lambda_n \mathscr{T}$ has scale $\lambda_n \to 0^+$. A continuum limit exists when there is a (domain independent) function $p(\lambda)$ such that the distribution of interfaces in $(\mathbb{D}_n, a_n, b_n)$ with parameters $p(\lambda_n)$ converges to some limit. The limiting value $p(0)$ is called the **critical value** and is denoted by p_c. The choice $p(\lambda) = p_c$ leads to a scale-invariant theory.

A map $f : \mathbb{D} \to \mathbb{D}'$ between two domains sending marked boundary points to marked boundary points (i.e. $f(a) = a'$ and $f(b) = b'$) is said to be **conformal** if it preserves angles. Riemann's theorem, to be explained in more detail in Sect. 2.3.1 asserts the existence of such maps. One can then ask, for a given interface model, whether the distribution of interfaces in $(\mathbb{D}, a, b)$ and in $(\mathbb{D}', a', b')$ are conformally equivalent. This can be checked numerically on good lattice approximations of these domains.

In two dimensions, scale-invariance plus locality is often enough to ensure conformal invariance. Thus the limiting theory of a discrete model at $p = p_c$ is a good candidate for conformal invariance. More generally, there is often a **threshold function** $p_s(\lambda)$ such that if $p(\lambda) - p_c = o(p_s(\lambda) - p_c)$ the limiting continuum theory is the same as the critical theory, if $p(\lambda) - p_c \sim p_s(\lambda) - p_c$ a limiting continuum theory exist but is not scale-invariant, and if $p_s(\lambda) - p_c = o(p(\lambda) - p_c)$ the limiting theory either does not exist or is trivial in some sense (concentrated on a single curve for instance). It is clear that only the small-λ behaviour of p_s matters and commonly $p_s(\lambda) - p_c$ can be taken to be simply a power of λ. The exponent is one of the critical exponents of the model.

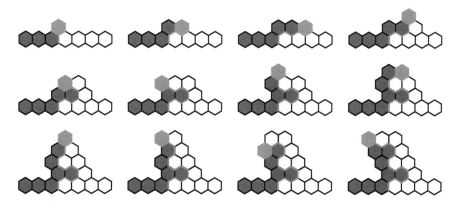

Fig. 2.4 The exploration process

2.2.2 The Exploration Process and Percolation

2.2.2.1 Definition of the Exploration Process

Let $(\mathbb{D}, a, b)$ be a hexagonal domain with admissible boundary condition. Colour the left hexagons in black (say) and the right hexagons in white. If a is incident to no inner hexagon of $\mathbb{D}$, all paths from a to b in $\mathbb{D}$ start with the same edge. Else, a is incident to exactly one inner hexagon of $\mathbb{D}$. Colour it black or white using a biased coin (say black has probability p and white $1 - p$), and make a step along the edge of $\mathbb{D}$ adjacent at a whose adjacent faces have different colours. Then remove from $\mathbb{D}$ the edge corresponding to the first step and its other end point, call it $\dot{a}$ to get a new domain $\dot{\mathbb{D}}$. If $\dot{a} = b$ stop. Else $(\dot{\mathbb{D}}, \dot{a}, b)$ is a new hexagonal domain with admissible boundary condition and one can iterate as shown in Fig. 2.4. Each choice of colour is made independently of the preceding ones but with the same bias. This random process is called the **exploration process**, and by construction it results in a simple path from a to b.

The fact that at some times the next step can be decided without tossing (for example, in Fig. 2.4, for the transition from the second picture to the third one, the choice of Colour for one hexagon is enough to fix two steps of the exploration path) results in a subtle interaction between the abstract independent coin tossings and their intricate effect on the geometry of the path.

There is exactly one coin toss for each inner face of $\mathbb{D}$ touching an edge of the path: this toss takes place the first time the inner face is touched by the tip of the path. In the rest of the process, this face becomes a boundary hexagon. But the path can have more than one edge along it.

The exploration process has a very important property: **locality**. It means that if $(\mathbb{D}, a, b)$ and $(\mathbb{D}', a, b)$, $\mathbb{D}' \subset \mathbb{D}$ are two hexagonal domains with the same admissible boundary condition (a, b), the distributions of the exploration path in $\mathbb{D}$ and in $\mathbb{D}'$ coincide up to the first time the exploration path touches a boundary hexagon of

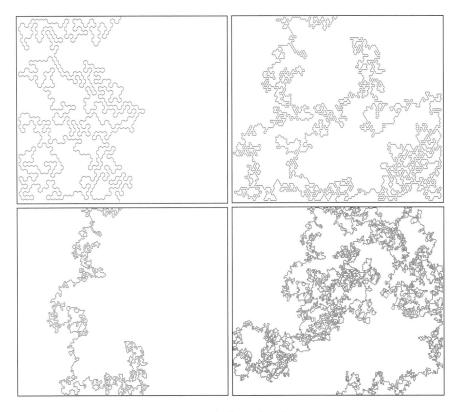

Fig. 2.5 Samples of the exploration process for increasing sizes

$\mathbb{D}'$ which is an inner hexagon of $\mathbb{D}$. This notion of locality should not be confused with the notion of locality in quantum field-theory.

By symmetry, if there is a single value of p for which the theory is critical, it has to be $p_c = 1/2$ and the numerics confirms this intuition. Figure 2.5 shows a few samples of the symmetric exploration process. They join the middle horizontal sides of similar rectangles of increasing size. The pseudo-random sequence is the same for the four samples.

Even for small samples, the exploration process makes many twists and turns. By construction, the interface is a simple curve, but with the resolution of the figure, the exploration process for large samples does not look like a simple curve at all!

To estimate the (Hausdorff, fractal) **dimension** of the symmetric exploration process, one can generate samples in similar rectangular domains of different sizes and made the statistics of the number of steps S of the interface as a function of the size L of the rectangle domain. One observes that $S \propto L^\delta$ and a modest numerical effort (a few hour of CPU) leads to $\delta = 1.75 \pm .01$. To get an idea of how small the finite size corrections are, observe Fig. 2.6.

The exploration process is build by applying local rules involving only a few nearby sites, and we could wave our hands to argue that its scale-invariance (for

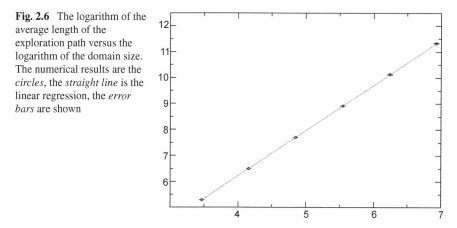

Fig. 2.6 The logarithm of the average length of the exploration path versus the logarithm of the domain size. The numerical results are the *circles*, the *straight line* is the linear regression, the *error bars* are shown

$p = p_c = 1/2$) should imply its conformal invariance in the continuum limit. Fortunately, hand-waving is not needed because *the exploration process* (on hexagonal domains) is one which *has been rigorously proved to have a conformally invariant distribution in the continuum limit, the fractal dimension being exactly* 7/4. As suggested by numerical simulations, the continuum limit does not describe simple curves but curves with a dense set of double points, and in fact the—to be defined later—SLE_6 process describes not only the growth of the exploration path but also the growth of the **exploration hull**, which is the complement of the set of points that can be joined to the end point by a continuous path that does not intersect the exploration path. As we shall explain in detail in Sect. 2.4.5, among SLE_κ's, SLE_6 is the only one that satisfies locality, so what is really to prove in this case is conformal invariance in the continuum limit (a non-trivial task), and the value of κ is for free.

2.2.2.2 Relation to Percolation and Coupling

The exploration process has been presented as a growth process, but in fact it is related to statistical mechanics in a simple way. Indeed, suppose that once the exploration sample has been constructed one tosses repeatedly (independently) the same coin to Colour also the hexagons that have not been coloured during the construction of the path. One gets a configuration in which all hexagons have been coloured independently, and from which the exploration path can be reconstructed has the sole curve joining the marked points with boarding hexagons all black one the left side and white on the right side. So one could also construct exploration samples by colouring all the hexagons independently at once and then drawing the separating curve. To summarise, *the exploration path is the interface for the statistical mechanics of* **percolation**.

Of course this approach is a poor idea for numerical simulations of the exploration process for a fixed p, because many hexagons are coloured for no use. But it has several advantages. First, it shows plainly that the law for the exploration process is reversible (i.e. the choice of which of the two marked is used to start the exploration path is irrelevant). Second, percolation can be studied with other boundary conditions. Third, it makes it possible to use the powerful probabilistic tool of *coupling*, to which we turn now.

It happens frequently that on some **measure space** $(A, \mathscr{F})$ one has to deal with a family of probability laws P_u where u is some parameter. Quite often the parameter u takes only two values, but this is not mandatory. In some favourable circumstances, one can find another **probability space** $(E, \mathscr{G}, \mu)$ and a family of measurable maps $f_u : E \to A$ such that the image measure of μ by f_u is P_u: if B is a measurable subset of A (i.e. $B \in \mathscr{F}$) then $f_u^{-1}(B)$ is a measurable subset of E (i.e. $f_u^{-1}(B) \in \mathscr{G}$) and $P_u(B) = \mu(f_u^{-1}(B))$. Thus we can fix a configuration in E and by changing u see a "movie" of configurations in A. This is known as **coupling**.

Abstractly, couplings always exist, but these general constructions are of little or no use. A coupling useful to tackle a given situation does not always exist, and even it does, it may take a good amount of creative skills to discover it.

However, in the case of percolation, it is easy to find a useful coupling. Let the parameter u vary in $[0, 1]$. If H is the set of inner hexagons (the ones whose colours are not fixed by boundary conditions) of some finite hexagonal domain, set $A = \{b, w\}^H$ with $\mathscr{F} = \mathscr{P}(A)$ (all subsets of A are measurable), and set $E = [0, 1]^H$ with μ the product Lebesgue measure. So A is the set of assignments of a colour, b(lack) or w(hite), to each inner hexagon, and E is the set of assignments of a real number $\in [0, 1]^H$ to each inner hexagon. A configuration in A can be seen equivalently as a map from H to $\{b, w\}$, or as a partition of H in black and white. If $x \in [0, 1]$, set $f_u(x) = b$ if $u < x$ and $f_u(x) = w$ if $u \geq x$. Use the product structure of A and E to define $f_u : E \to A$ so that an hexagon h is white if and only if its assigned value is $\geq u$. Obviously the image measure of μ by f_u colours the inner hexagons independently, each being black with probability $1 - u$ and white with probability u.

In such a setting, a useful tool is **Russo's formula**. Let us derive it abstractly and then interpret it. Suppose we partition A in two subsets $A = B \cup W$ in such a way that being in W is a so-called **increasing property**: *if $\gamma \in W$ and if $\gamma' \in A$ is such that all hexagons which are white in the configuration γ are also white in the configuration γ' then $\gamma' \in W$.* We order $\{b, w\}$ by saying that $w > b$ and use this to define a partial order: $\gamma' \geq \gamma$ if and only if all hexagons which are white in the configuration γ are also white in the configuration γ'. Viewing γ and γ' as maps from H to $\{b, w\}$, this says that if $\gamma \in W$ and $\gamma' \geq \gamma$ then $\gamma' \in W$. Then it is intuitively clear, and coupling makes it obvious, that $P_u(W)$ is an increasing function of u. If $\gamma \in A$ is a configuration, call a hexagon h **pivotal** for $A = B \cup W$ in the configuration γ either if $\gamma \in B$, h is coloured in black and changing it into white yields a configuration in W or if $\gamma \in W$, h is coloured in white and changing it into black yields a configuration in B. In the first case, we say that h is **pivotal** in γ to enter W and in the second case that h is pivotal in γ to enter B. In each configuration $\gamma \in A$ there is a certain number (possibly 0) of pivotal hexagons $\Pi(\gamma)$, and Π is thus a random variable on A. **Russo's formula** states that

$$\frac{\mathrm{d}}{\mathrm{d}u} P_u(W) = E_u(\Pi), \tag{2.2}$$

i.e. that the derivative of $P_u(W)$ is the expected number of pivotal points for the probability P_u.

Proof The proof is easy. We shall prove a slightly more refined identity, namely that if $\Pi_W(\gamma)$ is the number of hexagons in γ pivotal to enter W then $(1-u)\frac{\mathrm{d}}{\mathrm{d}u}P_u(W) = E_u(\Pi_W)$. By symmetry, if $\Pi_B(\gamma)$ is the number of hexagons in γ pivotal to enter B then $u\frac{\mathrm{d}}{\mathrm{d}u}P_u(W) = E_u(\Pi_B)$. As $\Pi = \Pi_W + \Pi_B$, the sum of these two equalities gives Russo's formula. Suppose $0 \le u < v \le 1$. By definition $P_v(W) - P_u(W) = \mu(f_v(X) \in W) - \mu(f_u(X) \in W)$ and by the increasing property of W this is $\mu(f_v(X) \in W$ and $f_u(X) \notin W)$. We can split this as a double sum to get

$$P_v(W) - P_u(W) = \sum_{\beta \in B}\sum_{\omega \in W} \mu\big(f_v(X) = \omega \text{ and } f_u(X) = \beta\big).$$

Note that the summand can be non-zero only if $\beta < \omega$ i.e. if one can go from β to ω by turning some black hexagons to white ones because this is what happens to $f.(X)$ for a fixed X by tuning the parameter from u to v. For a given X the hexagons h that change colour are those for which $X(h) \in]u, v]$, so in the above double sum only the pairs (β, ω) which disagree on a single hexagon can contribute to first order in $v - u$.

For instance we can sum first over β's to get

$$P_v(W) - P_u(W)$$
$$= \sum_{\beta \in B}\sum_{h \text{ pivotal in } \beta} \mu\big(f_u(X) = \beta \text{ and } X(h) \in]u, v]\big) + O\big((v-u)^2\big).$$

But by the definition of μ and f_u,

$$\mu\big(f_u(X) = \beta \text{ and } X(h) \in]u, v]\big) = \mu\big(f_u(X) = \beta\big)\frac{v-u}{1-u} = P_u(\beta)\frac{v-u}{1-u}.$$

In consequence,

$$P_v(W) - P_u(W) = \frac{v-u}{1-u}\sum_{\beta \in B} P_u(\beta)\#\{\text{pivotal points in } \beta\} + O\big((v-u)^2\big).$$

The sum is just the expected number of pivotal points to enter W for P_u, and taking the limit leads to the announced result:

$$(1-u)\frac{\mathrm{d}}{\mathrm{d}u}P_u(W) = E_u(\Pi_W).$$

Had we decided to sum first over ω's, we would have obtained

$$u\frac{\mathrm{d}}{\mathrm{d}u}P_u(W) = E_u(\Pi_B). \qquad \square$$

Now that we have proved Russo's formula abstractly, let us apply it to a concrete decomposition of A relevant for percolation. Take a domain and split its boundary into four segments, such that the colours of the hexagons are fixed on each segment but alternate from one segment to the next as in Fig. 2.7. Then a simple topological argument shows that in any configuration either there is a black cluster connecting

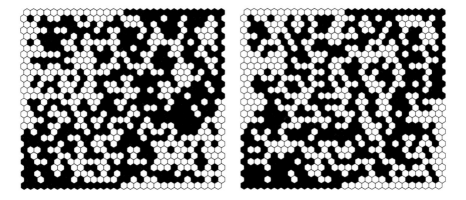

Fig. 2.7 *Left panel*: a sample of critical percolation with a black crossing and no pivotal point. *Right panel*: a sample of critical percolation with a white crossing and several pivotal points. Find these!

the two black boundary components, or there is a white cluster connecting the two white boundary components. In the first case put the configuration in B and in the second case put it in W. That being in W is an increasing property is clear. **Pivotal hexagons** are the ones which change the colour of the connecting cluster, so they have a impact on the long range properties of a configuration. Figure 2.7 shows two samples.

Such pivotal points could be called "global pivotal points" because they are defined with respect to global boundary conditions. However, in an arbitrary configuration of percolation one can look at windows of a certain size and define pivotal points with respect to that window. Anyway, at $u = p_c = 1/2$, the number of pivotal points can be shown to behave like $(L/\lambda)^{3/4}$ where L is the linear size of the system and λ is the scale of the tiling. So $p_s(\lambda) - 1/2 = \lambda^{3/4}$ is a good candidate for the threshold function: by coupling, changing u from the critical value $1/2$ to $1/2 + g\lambda^{3/4}$ just flips of order 1 pivotal points, and Russo's formula indicates that $P_{1/2+g\lambda^{3/4}}(W) - P_{1/2}(W)$ is a finite function of g in the continuum limit. The validity of this threshold function is in fact rigorously proved (only in some weak sense at the moment, but progress is rapid).

At that point, an instructive subtlety enters the game. As physicists, we expect that a continuum field-theory describing the vicinity of the critical point exists. This theory will depend on a renormalised parameter that can be taken to be g or a correlation length. We also expect that correlation functions (involving a finite number of points) in this theory will depend smoothly on g. However, on can show in this example that in the case of percolation, *the probability measure on the interface, which is even a more global observable, does not depend smoothly on g.*

Before we consider percolation, which is already complicated, we take a glance at a simpler case, $1D$ random walks. This will allow us to explain the intuitive meaning of the smoothness statement.

Let λ be the lattice spacing, p be the probability to make a step to the right and $1 - p$ the probability to make a step to the left. Suppose we make n steps. So a walk starting at point x has its endpoint at $x + S_n$ where $S_n = \lambda(\varepsilon_1 + \ldots + \varepsilon_n)$. The random variables $\varepsilon_1, \ldots, \varepsilon_n$ are independent and take value 1 or -1 with probability p or $1 - p$ respectively. It is clear that coupling can be used in this case and an analysis analogous to percolation could be carried out. One way to do so would be to consider an interval of length L containing the origin and take $S_0 = x \in [0, L]$. Then $P = P(p, \lambda, x, L)$, the probability that the first exit of $]0, L[$ is at L, plays the role of a **crossing probability** in percolation. The analysis is elementary but as the outcome is well-known, we do not go into the details. The salient features are that $p_c = 1/2$ is the critical value, that the number n of steps to exit the interval is $\sim (L/\lambda)^2$ and that $p_s = 1/2 + \lambda$ is a threshold function.

For given p, the mean (expected value) of S_n, which measures the asymmetry between steps to the right and steps to the left, is $\lambda n(2p - 1)$ and the fluctuation (variance) of S_n is $2\lambda\sqrt{np(1 - p)}$. When the fluctuation is much smaller that the mean, a look at a sample will be enough evidence to decide with overwhelming confidence that it was not drawn with a symmetric distribution. However, if the asymmetry and its fluctuations are of the same order, one cannot know for sure. In the continuum limit, setting $p = 1/2 + g\lambda$ and $n \sim (L/\lambda)^2$, we find that the mean is $\sim L^2 g$ and that the fluctuation is $\sim L$. Both these quantities are finite for $\lambda \to 0^+$, so it is impossible to decide whether a sample long enough to exit the interval was drawn with the symmetric distribution or not. A typical sample for the symmetric distribution is also typical for the asymmetric continuum distribution. Another way to check this statement is to look at a symmetric sample of n steps. Its probability is $(1/2)^n$. Now for the asymmetric distribution, the same sample has probability $(p(1 - p))^{n/2}(p/(1 - p))^{S_n/2\lambda}$. The ratio, which quantifies how less likely the symmetric sample is under the asymmetric distribution, is $(4p(1 - p))^{n/2}(p/(1 - p))^{S_n/2\lambda}$. Again, if $p = 1/2 + g\lambda$ and $n \sim (L/\lambda)^2$, the ratio is finite when $\lambda \to 0^+$, a fact which can be seen as a (weak) form of the statement that the ratio between the symmetric distribution and the asymmetric continuum distribution is finite: the continuum distribution depends smoothly on g.

Now we can go back to the case of percolation. We shall give first a very crude argument and then a crude one. A rigorous proof is really involved.

The typical fluctuation of the number of occupied sites in a percolation sample is $\sim L/\lambda$ because there are $\sim (L/\lambda)^2$ independent sites. However if $p = 1/2 + g\lambda^{3/4}$, the typical asymmetry is $\sim g\lambda^{3/4}(L/\lambda)^2 = gL^{3/4}(L/\lambda)^{5/4}$, which is much larger that the fluctuation $\sim L/\lambda$ when $\lambda \to 0^+$, so one can assert with certainty that an individual sample is critical or not. In fact, the same counting implies that on any set containing $\sim (L/\lambda)^d$ hexagons and chosen independently of the percolation sample, the asymmetry $\sim g\lambda^{3/4}(L/\lambda)^d$ is much larger than the fluctuation $\sim (L/\lambda)^{d/2}$ as soon as $d > 3/2$. The critical percolation interface is bounded by $\sim (L/\lambda)^{7/4}$ hexagons so it covers enough of the sample to feel a macroscopic effect of the tiny bias out of criticality. Of course, this is cheating because the interface as a set is correlated to the hexagon configuration.

To do a bit better, we need a bit more knowledge. Take an interface drawn from the symmetric distribution. We view the interface as the outcome of an exploration process. Let n be the total number of choices. It is intuitive, and can be proved, that n

scales like the total length of the interface, i.e. for a typical symmetric interface $n \sim$ $(L/\lambda)^{7/4}$. If d is the difference between the number of black and white choices, i.e. the asymmetry, the usual rules of chance ensure that $d \sim n^{1/2}$. So $d \sim (L/\lambda)^{7/8}$. For an asymmetric distribution, the mean asymmetry for n choices is of order $n(2p-1)$, which for $p = 1/2 + g\lambda^{3/4}$ yields $\sim gL^{7/4}\lambda^{-1}$. This is much larger than $(L/\lambda)^{7/8}$ when $\lambda \to 0^+$. Hence an interface which is typical for the symmetric distribution is very atypical for an asymmetric continuum distribution.

2.2.2.3 General Remarks on Interfaces

We would like to extract some general features of interfaces that go beyond percolation. The crucial observation is the following. We have defined percolation configurations with an interface for any hexagonal domain $(\mathbb{D}, a, b)$ with admissible boundary condition. For percolation, the weight of a configuration is a product of independent weights, and partition functions are always 1. But the existence of an interface is a "topological" fact: whatever the Boltzmann weights given to a colouring of the inner hexagons in $(\mathbb{D}, a, b)$, the colouring will allow to identify a unique well-defined interface connecting a and b.

To define models generalising percolation, a crucial step is to see how the information encoded in hexagon colourings can be retrieved from another type of geometric data.

For each configuration, one paints the edges separating hexagons of different colours. In particular, the edges forming the interface are among the painted edges, but there are many more painted edges in general. The topology of the hexagonal lattice leads to interesting consequences. Consider a vertex not on the boundary. It touches three edges and three hexagons. By direct enumeration of the possible colourings of these three hexagons, one checks that either 0 or 2 painted edges contain the vertex. A boundary vertex touching only two hexagons can belong to 0 or 1 painted edge. A boundary vertex touching only one hexagon can belong to no painted edge. Consequently, if the painted edges are grouped in connected components, these components are simple curves (no branchings) that can only end at a boundary vertex. For admissible boundary conditions, this implies that apart from the interface, which ends at a and b, all the other curves are closed loops, and the painted edges form a gas of self avoiding loops, plus the interface. An example is shown in Fig. 2.8. Of course, would one fix all boundary hexagons to be of the same colour, one would have no interface but only a loop gas. At the other extreme, if the boundary conditions would allow many colour changes, one would see a gas of interfaces and loops. In any case, the curves split the domain in connected components in which all hexagons have the same colour. On the other hand, by definition, each time a curve is crossed, the hexagon colour changes. So one can proceed backwards: if one keeps the curves but erases the hexagon colours except for a single hexagon, the full colouring can be retrieved. If all hexagon colours are erased, there is a twofold degeneracy in the reconstruction of hexagon colours from the curve configuration. Of course, depending on the boundary conditions on hexagon colours, only certain curve arrangements can occur.

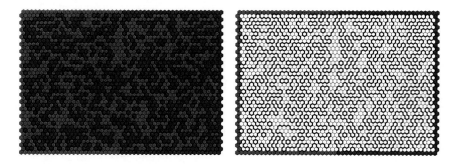

Fig. 2.8 A critical percolation sample with the hexagons colours and loops (*left*) and the same configuration with only the loops (*right*)

The **loop description** is useful for several reasons. The first one is that it leads to popular statistical mechanics models. The second is that it puts the study of the interface on some natural footing. Indeed, the loops and the interface are objects of the same nature. In particular from a physical viewpoint they are expected to share many local properties. This is especially true at a critical point when comparing large loops and the interface.

There are simple ways to associate a Boltzmann weight to a loop (or loop + interfaces) configuration. A **configuration** ℓ is made of a number of **edges**, say E_ℓ, these edges building a number, say C_ℓ, of connected components (i.e. of disjoints simple curves). We can then define a **Boltzmann weight** depending on 2 parameters, K and n by the formula

$$w(\ell) = K^{E_\ell} n^{C_\ell}. \tag{2.3}$$

Note that with this weight, the two-fold degeneracy between the hexagon colour configuration and the loop (or loop + interfaces) description is irrelevant. The parameter K is a kind of fugacity, and n (if an integer) is related to some group theory factors in another formulation of the model. Though we shall not explain why here, this is the reason why one talks generically of the **O(n) model**. **Percolation** is described by $n = 1$ and $K = 1$. Keeping $n = 1$, K can be seen as $\tan \beta$ (i.e. as a temperature-like parameter) for a very famous model, the Ising model,[2] as shown by a standard large temperature expansion. So there is another value of K, namely $K = 3^{-1/2}$ for which the model is again critical. A glance at Fig. 2.9, compared to Fig. 2.8 suggests that percolation leads to denser loops than the critical Ising model. This is a general feature: for $n \in [-2, 2]$ by adjusting $K = K_c^- := (2 - (2 - n)^{1/2})^{-1/2}$ one gets a critical dense loop gas and by adjusting $K = K_c^+ := (2 + (2 - n)^{1/2})^{-1/2}$ one gets a critical dilute loop gas. The section on loop-erased random walks shows that they are closely related to $n = -2$ in the dilute phase. Other values of n describe other systems of interest. For instance $n = 2$

---

[2]Warning: this relationship between percolation and the Ising model is special to hexagonal domains!

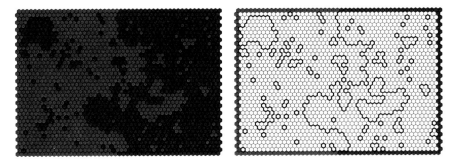

Fig. 2.9 An Ising sample with the hexagons colours and loops (*left*) and the same configuration with only the loops (*right*). The sample is only "close to critical" due to finite size effects

is related to the XY model, the Kosterlitz-Thouless transition and the Gaussian free field, $n = 0$ to self-avoiding walks (obvious with our definitions) and so on.

Let us close this short presentation by a few remarks. Keeping the domain fixed, let Z_{ab} denote the partition in $(\mathbb{D}, a, b)$ with admissible boundary conditions, and Z the same partition but with all boundary hexagons fixed to be of the same colour. The ratio Z_{ab}/Z bears some analogy with a 2-point correlation function. Observe that the inner hexagon colour configurations that contribute to Z_{ab} and Z are the same, but the loop configurations are very different. Of course the loops that do not touch the boundary hexagons are the same in both cases, but the loops that touch the boundary are sensitive to a change of boundary conditions. For analogous reasons, in general, there is no observable O_a such that $Z_{ab} = \sum_\ell w(\ell) O_a(\ell) O_b(\ell)$. However, one can find observables O_a^i, indexed by some parameter i such that $Z_{ab} = \sum_\ell w(\ell) \sum_i O_a^i(\ell) O_b^i(\ell)$. This may seem artificial, but is in fact closely related to the channels that appear in a quantum field-theory operator description of the statistical system when Z_{ab}/Z is identified with a vacuum expectation value $\langle \Omega | \psi(a) \psi(b) | \Omega \rangle$. Note also that if one attributes weight 1 instead of n to loops that touch the boundary, one can locally find observables O_a that allow to interpret Z_{ab}/Z as an almost *bona fide* statistical 2-point correlation function.

2.2.3 Loop-Erased Random Walks

This example still keeps some aspects of a growth process, in that new pieces of the process can be added recursively. Loop-erased random walks were invented by Lawler as an example of random paths more tractable than the canonical self avoiding walks. A **loop-erased random walk** (LERW) is a random walk with loops erased along as they appear.

2.2.3.1 Definition

More formally, if $X_0, X_1, \ldots, X_n$ is a finite sequence of abstract objects, we define the associated **loop-erased sequence** by the following recursive algorithm.

```
Initialise counters l = 0 and m = 1 and set Y₀ = X₀
Iterate while m ≤ n
{
    - If there is a k with 0 ≤ k ≤ l such that Yₖ = Xₘ set l = k
    - Else increment l by 1 and set Yₗ = Xₘ.
}
The loop-erased sequence is Y₀,..., Yₗ.
```

Let us look at two examples.

1. For the "month sequence" $j, f, m, a, m, j, j, a, s, o, n, d$, the first loop is m, a, m, whose removal leads to $j, f, m, j, j, a, s, o, n, d$, then j, f, m, j, leading to j, j, a, s, o, n, d, then j, j leading to j, a, s, o, n, d where all terms are distinct.

2. For the "reverse month sequence" $d, n, o, s, a, j, j, m, a, m, f, j$, the first loop is j, j, leading to $d, n, o, s, a, j, m, a, m, f, j$, then a, j, m, a leading to d, n, o, s, a, m, f, j.

This shows that the procedure is not "time-reversal" invariant. Moreover, terms that are within a loop can survive: in the second example m, f, which stands in the j, m, a, m, f, j loop, survives because the first j is inside the loop a, j, m, a which is removed first.

The above algorithm is most useful if the sequence $X_0, X_1, \ldots, X_n$ is viewed as a stream of data that is treated "on the fly". If $X_0, X_1, \ldots, X_n$ is known at once, another algorithm erases the loop in possibly fewer steps. It goes as follows:

```
Initialise counters l = 0 and m = n
Until l = m, iterate
{
    - Find the largest k ≤ m such that Xₖ = Xₗ
    - If k > l remove the terms with indices from l + 1 to k, and
        shift the indices larger than k by l − k to get a new
        sequence.
    - Decrement m by k − l and increment l by 1.
}
```

For the "month sequence", this leads at once from $j, f, m, a, m, j, j, a, s, o, n, d$ to j, a, s, o, n, d, and then the counter l is incremented from 0 to 5 without further removals. For the "reverse month sequence", the counter l is incremented from 0 to 4, a loop is removed leading from $d, n, o, s, a, j, j, m, a, m, f, j$ to d, n, o, s, a, m, f, j, then the counter l is incremented from 5 to 7 without further removals.

Fig. 2.10 A loop-erased
random walk with its shadow

A **loop-erased random walk** arises when this procedure is applied to a (two-dimensional for our main interest) random walk. In the full plane this is very easy to do. Figure 2.10 represents a loop-erased walk of 200 steps obtained by removing the loops of a 4006 steps random walk on the square lattice. The thin grey lines build the shadow of the random walk (where **shadow** means that one does not keep track of the order and multiplicity of the visits) and the thick line is the corresponding loop-erased walk. The time-asymmetry is clearly visible and allows to assert with little uncertainty that the walk starts on the top right corner.

In this setting, it is trivial to get samples but the measure remains in the background. One possible approach is the following. Consider a symmetric random walk on the square lattice and view the successive positions as a stream of data. Remove the loops as they show up, and stop the random walk at the first time n for which the associated loop-erased walk has reached length N. The probability of the random walk is 4^{-n}. Note that the set of random walks for which the loop-erasure never reaches size N has probability 0, for instance as a subset of the set of random walks that remain in the ball of radius N centred at the origin forever. So the total probability for the set of random walks stopped when their loop-erasure reaches length N is 1. This procedure leads to a finite family of loop-erased walks, each of them can be obtained via the loop erasure of an infinite number of random walks. The probability of a given loop-erased walk is taken to be sum of the individual probabilities of its random walk ancestors.

This can be adapted to the setting of discrete domains with admissible boundary condition. Let $(\mathbb{D}, a, b)$ be such a domain, and let ν be the **coordination number** of the associated tiling, i.e. the number of edges adjacent to a vertex: ν is 6 or the triangular tilings, 4 for the square tilings, and 3 for the hexagonal tilings. Consider the set of walks from a to b in $\mathbb{D} \cup \{a, b\}$ that visit a and b only once, and give each step weight ν^{-1}, so that the weight of a walk is the usual random walk probability on the tiling. However, the total weight of walks from a to b in $\mathbb{D} \cup \{a, b\}$ is < 1. As before, these walks can be loop-erased, and the weight of a simple path γ from a to b in $\mathbb{D} \cup \{a, b\}$ is taken to be sum of the weights of all random walks from a to b in $\mathbb{D} \cup \{a, b\}$ whose loop erasure leads to γ. To get a probability measure, one needs to divide the weight by the total weight of all random walks from a to b in $\mathbb{D} \cup \{a, b\}$. This is easy in principle, and is closely related to the solution of the discrete Laplace equation with appropriate boundary conditions; in practice this normalisation can be computed explicitly for only a handful of examples.

In the same spirit, if we have an arbitrary weight assignment for walks from a to b in $\mathbb{D} \cup \{a, b\}$, we can use it to induce a weight on simple paths from a to b in $\mathbb{D} \cup \{a, b\}$ again by taking the weight of a simple path γ to be sum of the weights of all random walks from a to b in $\mathbb{D} \cup \{a, b\}$ whose loop erasure leads to γ. What is special about the standard random walk weight is that, as is well known, the random walk has a scale-invariant limit (Brownian motion of course), so the corresponding loop-erased random walk can be expected to have a scale-invariant limit. The loop-erased random walk is one of the first systems that has been proved to have a (not only scale—but even) conformally invariant continuum limit, the fractal dimension being $5/4$. A naive idea to get directly a continuum limit representation of loop-erased walks would be to remove the loops from a Brownian motion. This turns out to be impossible due to the proliferation of overlapping loops of small scale. However, the SLE_2 process, to be defined later, gives a direct definition. In fact, it is the consideration of loop-erased random walks that led Schramm to propose SLE as a description of interfaces.

2.2.3.2 Simulation

We have seen that is very simple to generate loop-erased random walks of a fixed length N in the plane. We could use this technique to get a probability measure on the first N steps of loop-erased random walks of length M. However, it is unclear whether this probability measure stabilises if we fix N and let M go to infinity. One of the problems is that in two dimensions, random walks are **recurrent**: with probability one they visit every site (and then they have to do it infinitely many times). So if we erase the loops of a random walk, the resulting loop-erased walk never stabilises; if we wait long enough, the random walk comes back to the origin and at that instant the loop-erased walk starts anew from scratch.

The numerical simulation of a loop-erased random walk in domains $(\mathbb{D}, a, b)$ is not easy either, because the random walks have a tendency to leave $\mathbb{D}$. Note that it would bias the sampling if we would forbid them to leave by simply dispatching the weight of steps leaving $\mathbb{D}$ to the ones staying in $\mathbb{D}$. What one has to do is to *condition* on random walks staying in $\mathbb{D}$. So most samples would have simply to be rejected and only from time to time would a sample be a walk from a to b in $\mathbb{D} \cup \{a, b\}$.

There is one exceptional domain in which at the same time an infinite loop-erased random walk can easily be defined and simulated. It is when $\mathbb{D}$ is the square tiling of a half-space, conventionally taken to be $\mathbb{H}_{\text{int}}$, the tiling of the upper-half-plane with vertices at the points $(n, m) \in \mathbb{Z} \times \mathbb{N}$, a is $O := (0, 0)$ (by translation-invariance along the real axis, any boundary vertex would do) and b is infinity. Let us explain why random walks on the square lattice conditioned to go from O to infinity while staying in $\mathbb{H}_{\text{int}}$ have a simple description.

The horizontal steps are not an issue, and we can concentrate on vertical steps. For a simple random walk in one dimension, it is well-known that a walk started at $m \in [0, l]$ touches the boundary for the first time at the endpoint 0 with probability $1 - m/l$ and at the endpoint l with probability m/l. Indeed, if $p(m)$ is the probability to touch the boundary for the first time at the endpoint 0, then $p(0) = 1$, $p(l) = 0$ and if $m \in]0, l[$, $p(m) = \frac{1}{2}(p(m - 1) + p(m + 1))$ as can be seen by conditioning on the first step of the walk. So by the usual rules of conditional probabilities, if the random walk is conditioned to exit at l and is at $m \in]0, l[$ at time t, it has probability $\frac{m+1}{2m}$ to go to $m + 1$ and $\frac{m-1}{2m}$ to go to $m - 1$. This has three striking consequences. First, the process remains Markov and time homogeneous. Second, the transition probabilities do not depend on l, so they can be used even if l is infinite. Third, taking l infinite, the probability, starting at m, never to visit $m' < m$ is $1 - m'/m > 0$ as can be seen by conditioning on the first step of the walk.

These three properties imply that each site is visited only a finite number of times, i.e. the walk escapes to infinity. Let us explain this in more detail. Suppose the walk starts from 0, goes to 1 at the first step, and then the above transition probabilities are used. Then at the second step the walk goes to 2. With probability $1/2$ it *never* goes back to 1 again. With probability $1/2$ it comes back to 1 at some point, and then the walk starts anew. Thus the number of visits to 1 follows a geometric law: 1 is visited $k \geq 1$ times with probability $1/2^k$. In particular the probability to visit 1 at least k times is $1/2^{k-1}$ which goes to 0 (in fact exponentially). Hence with probability 1 the number of visits of point 1 is finite. The same argument generalises. First, let s be the probability that point $m \geq 1$ is never visited. Suppose the walk is at point m. With probability $\frac{m+1}{2m}$ it goes to $m + 1$ and then with probability $1 - \frac{m}{m+1} = \frac{1}{m+1}$ it *never* goes back to m again. So the total probability that the walk starting from m never visits m again is $r \geq \frac{1}{2m} > 0$. It follows that the number of visits to m follows essentially a geometric law: m is visited 0 times with probability s and $k \geq 1$ times with probability $(1 - s)r(1 - r)^{k-1}$. Again, the probability to visit m at least $k \geq 1$ times is $(1 - s)(1 - r)^{k-1}$ which goes to 0 (in fact exponentially). Hence with probability 1 the number of visits of point m is finite. This is true for any m and any starting point for the walk. Hence, in particular we see recursively that if the walk is at $m' < m$, m will be visited later with probability one, because with probability 1 all points in $[0, m - 1]$ are visited only finitely many times. This means that in fact $r = \frac{1}{2m}$. To summarise, if the walk starts from 0, the number of visits to $m \geq 1$ is $k \geq 1$ with probability $\frac{1}{2m}(1 - \frac{1}{2m})^{k-1}$. In particular, the walk is transient, i.e. it escapes to ∞ with probability 1.

So we use the usual random walk in the horizontal direction but the conditioned random walk in the vertical direction. Explicitly, at the first step the walk goes from $(0, 0)$ to $(-1, 1)$ or $(1, 1)$ with probability $1/2$, and later, if at (n, m), the walk makes a step in the NE or NW directions with probability $\frac{m+1}{4m}$ and in the SE or SW direc-

Fig. 2.11 A sample of the loop-erased random walk

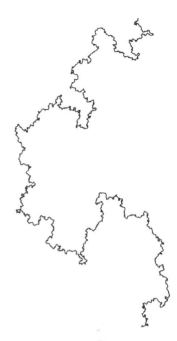

tions with probability $\frac{m-1}{4m}$. Call the vertical coordinate **altitude** for convenience. As explained before, the altitude of the walk goes to ∞ with probability one, and the associated loop-erased walk converges. More precisely, for $m < M$ stop the random walk the first time it reaches altitude M and stop the corresponding loop-erased random walk at altitude m. Then with probability $> 1 - m/M$ the loop-erased random walk up to altitude m will not be modified by the subsequent evolution of the random walk. This is because to close a loop, the walk has to come back to the same point, which is more stringent than to come back to the same altitude. Hence, letting M go to infinity, we get a well defined limiting distribution for loop-erased random walks from O to altitude m for any m, hence for loop-erased random walks from O to ∞. Accurate numerical simulations are made by taking $M \gg m$. However, the process for which $m = M$ is interesting as well. It has a continuum limit which can be studied with the so-called dipolar variant of stochastic Loewner evolutions.

Figure 2.11 shows a sample of loop-erased random walk of about 10^5 steps. At first glance, one observes a simple (no multiple points) irregular curve with a fractal structure. The intuitive explanation why a loop-erased random walk has a tendency not to come back too close to itself is that if it would do so, then with large probability a few more steps of the random walk would close a loop.

2.2.3.3 Relation with Statistical Mechanics

Again, it is useful to make the connection between the purely geometric description of loop-erased random walks and more conventional statistical mechanics.

The starting point of the correspondence is a formula for the expansion of the determinant $\det(\mathbf{1} - \mathbf{A})$ when $\mathbf{A} = (A_{vv'})_{v,v' \in V}$ is a matrix with index set V. For later convenience, we call the elements of V **vertices**. A **cycle** of length $k \geq 1$ in V is sequence $(v_1, \ldots, v_k)$ of distinct vertices of V modulo cyclic permutation; so that $(v_1, \ldots, v_k)$, $(v_2, \ldots, v_k, v_1)$, $\ldots$ represent one and the same cycle. Cycles are said to be **disjoint** if no vertex appears in more than one of them. The subsets $\{C_1, \ldots, C_n\}$ of $\mathscr{P}(V)$ made of n disjoint cycles of V form a set that we denote $\mathscr{C}_n$. The **weight** of a cycle C represented by $(v_1, \ldots, v_k)$ is by definition $w(C) := A_{v_1 v_2} \ldots A_{v_{k-1} v_k} A_{v_k v_1}$ (for $k = 1$, this reduces to $A_{v_1 v_1}$) which indeed is invariant under cyclic permutations.

An elementary reorganisation of Cramer's formula yields

$$\det(\mathbf{1} - \mathbf{A}) = \sum_{n \geq 0} (-1)^n \sum_{\{C_1, \ldots, C_n\} \in \mathscr{C}_n} w(C_1) \ldots w(C_n). \tag{2.4}$$

Similarly, for $v, v' \in V$ we define a **walk** of k steps from v to v' in V to be any sequence of vertices $(v_0, \ldots, v_k)$ with $v_0 = v$ and $v_k = v'$ but with $v_1, \ldots, v_{k-1}$ distinct from v and v'. Hence with this definition a walk visits its starting and end point only once. This restriction is a bit unusual, but it is not really crucial. The weight of a walk $W = (v_0, \ldots, v_k)$ is taken to be $w(W) := A_{v_0 v_1} \ldots A_{v_{k-1} v_k}$.

The sequence W can be loop-erased to yield a path from v to v' (remember that paths are walks in which a given vertex appears at most once). If γ is a path, we define

$$\tilde{w}(\gamma) := \sum_{W \mapsto \gamma} w(W), \tag{2.5}$$

where the sum is over all walks whose associated loop-erased walk is γ. We aim at a general formula for $\tilde{w}$.

Let $\gamma = (v_0, \ldots, v_k)$ be a path from $v, v' \in V$. Let $V^{(0)} := V \backslash \{v_k, v_0\}$, $V^{(1)} := V \backslash \{v_k, v_0, v_1\}, \ldots$. For $l = 0, \ldots, k-1$ let $\mathbf{A}^{(l)}$ be the matrix $\mathbf{A}$ restricted to the vertex set $V^{(l)}$.

A walk W which yields γ after loops have been erased can be decomposed as follows (see the second loop-erasing algorithm on p. 67): the walk (v_0, v_1), followed by an arbitrary number of walks from v_1 to v_1 in $V^{(0)}$, followed by the walk (v_1, v_2), followed by an arbitrary number of walks from v_2 to v_2 in $V^{(1)}$ and so on. Take $1 \leq l \leq k-1$. Note that if one expands $(\frac{1}{1-\mathbf{A}^{(l-1)}})_{v_l v_l}$ in formal power series in the coefficients of $\mathbf{A}$, one gets exactly the sum of the weights for the concatenation of an arbitrary number of walks from v_l to v_l in $V^{(l-1)}$. Hence we infer that

$$\tilde{w}(\gamma) = A_{v_0, v_1} \left(\frac{1}{1 - \mathbf{A}^{(0)}} \right)_{v_1 v_1} A_{v_1, v_2} \left(\frac{1}{1 - \mathbf{A}^{(1)}} \right)_{v_2 v_2} A_{v_2, v_3}$$
$$\times \ldots \times A_{v_{k-2}, v_{k-1}} \left(\frac{1}{1 - \mathbf{A}^{(k-2)}} \right)_{v_{k-1} v_{k-1}} A_{v_{k-1}, v_k}.$$

But by Cramer's formula for the inverse of a matrix,

$$\left(\frac{1}{1 - \mathbf{A}^{(l-1)}}\right)_{v_l v_l} = \frac{\det(1 - \mathbf{A}^{(l)})}{\det(1 - \mathbf{A}^{(l-1)})} \quad \text{for } l = 0, \dots, k-1.$$

Hence the product in the above formula for $\tilde{w}(\gamma)$ is telescopic, and we get the representation we were aiming at:

$$\tilde{w}(\gamma) = w(\gamma) \frac{\det(1 - \mathbf{A}^{(k-1)})}{\det(1 - \mathbf{A}^{(0)})}. \tag{2.6}$$

A first use of this formula is that is shows clearly that if the matrix $\mathbf{A}$ is symmetric, the loop-erased random walk weight is reversible i.e. the same for a path and its opposite or time reversal. In all cases the asymmetry comes solely from the weight of γ.

It is time to interpret the formulæ obtained so far in connection with statistical mechanics.

We start with Eq. (2.4) but read from right to left. The right-hand side can be seen as a partition function for a gas of oriented loops on a graph. Indeed, if E is an arbitrary subset of $V \times V$, we can consider the corresponding oriented graph $G = (V, E)$ i.e. view E as the set of edges if G. We give each edge in $(v, v') \in E$ the weight $A_{vv'}$ and impose that $A_{vv'} = 0$ if $(v, v') \notin E$. An **oriented loop** on G is a sequence $(v_1, \dots, v_k)$ of distinct vertices of V modulo cyclic permutation, with the condition that $(v_1, v_2), (v_2, v_3), \dots, (v_{k-1}, v_k), (v_k, v_1)$ are in E. Except for the last condition, this is what we called a cycle before: note that "cycle" reminds of the permutation context whereas "loop" reminds of geometric context. A configuration is a family of disjoint oriented loops, each oriented loop counts for a weight which is the product of the weight of the traversed edges and an overall factor (-1). Then the partition function, i.e. the sum of the weights of all possible configuration is by definition the right-hand side of Eq. (2.4), and this reconstructs the determinant on the left-hand side. We can specialise more by assuming further that E is a symmetric subset of $V \times V$ that does not meet the diagonal, and that $\mathbf{A}$ is symmetric. Then there is no loop of length 1, and the loop $(v_1, \dots, v_k)$ has the same weight as the loop traversed in the opposite order $(v_k, \dots, v_1)$. If $k = 2$ a loop and its opposite are the same, but not if $k \leq 3$. So we get the same partition function if instead of summing over oriented loops, we sum over un-oriented loops counting each un-oriented loop of length ≥ 3 twice, i.e. giving un-oriented loop of length ≥ 3 an overall factor (-2) instead of (-1). Finally, we could also give each edge in E the same weight K so that the weight of a loop configuration would be

$$(-1)^{\#\text{loops of length 2}} (-2)^{\#\text{loops of length} \geq 3} K^{\#\text{traversed edges}}$$

where of course loops of length 2 count for 2 traversed edges.

This statistical weight could be used as a definition of the so-called $\mathbf{O}(-2)$ **model**, where -2 reminds of the overall weight of each loop (of length ≥ 3). This model has several avatars, which are supposed to be in the same universality class, i.e. to describe the same macroscopic physics in the continuum limit. In certain versions, loops of length 2 are completely forbidden, at the price of renormalising K.

Replacing the factor (-2) by a factor n yields the general $O(n)$ model, which we have seen before.

Note that the partition function, i.e. $\det(1 - A)$ has a simple "field-theory" interpretation: if χ_v and $\bar{\chi}_v$, $v \in V$ are a collection of independent Grassmann variables, the fundamental result of Grassmann integration is

$$\det(1 - A) = \int \prod_{v \in V} \mathrm{d}\chi_v \, \mathrm{d}\bar{\chi}_v \, e^{\sum_{v,v'} \chi_v (\delta_{vv'} - A_{vv'}) \bar{\chi}_{v'}}.$$

This is the clue to the quantum field-theory approach to loop-erased random walks.

Before we interpret Eq. (2.6), let us start with a general observation. Suppose $\mathscr{C}$ is a configuration space, assumed to be finite for simplicity and consider a model of statistical mechanics on $\mathscr{C}$. Each $c \in \mathscr{C}$ has a weight $w(c)$. The **partition function** is $Z := \sum_{c \in \mathscr{C}} w(c)$. Suppose $\mathscr{C}$ can be partitioned as $\mathscr{C} = \bigcup_{\gamma \in \Gamma} \mathscr{C}_\gamma$. Then we can define $Z_\gamma := \sum_{c \in \mathscr{C}_\gamma} w(c)$ for $\gamma \in \Gamma$, and Z_γ can be interpreted as the marginal weight of $\mathscr{C}_\gamma$. The probability of $\mathscr{C}_\gamma$ is simply Z_γ/Z. In concrete situations, the splitting $\mathscr{C} = \bigcup_{\gamma \in \Gamma} \mathscr{C}_\gamma$ will usually have some interpretation. For instance, in the cases we are interested in these notes, we shall look at configuration spaces $\mathscr{C}$ that describe a statistical mechanics model on domains $(\mathbb{D}, a, b)$ with boundary conditions, in such a way that in each $c \in \mathscr{C}$ we can identify unambiguously a path γ joining a to b. Of course γ depends on c, and we can use this γ to split c. Then Z_γ/Z is simply the probability to observe the path γ. The reader should have another glance at Sect. 2.2.2.2 to look at the relationship between the exploration path and percolation from this viewpoint.

Equation (2.6) can then be interpreted straightforwardly. We consider now configurations made not simply of (mutually avoiding) loops, but of (mutually avoiding) loops avoiding a path from v to v'. The total weight of configurations for a fixed path from v to v' is simply the numerator of the right-hand side of Eq. (2.6). The denominator depends on v and v' but not on the simple path between them. So from the point of view of statistical mechanics explained before, the weight the loop-erased random walk model assigns to a path γ, i.e. the left-hand side of Eq. (2.6), is proportional to the marginal weight of configurations of "loops plus that path" in the loop gas model.

Hence we have succeeded in giving an interpretation of the loop erased random walks as interfaces in a statistical mechanics model. We are cheating a bit here because even if we take a positive edge weight K, due to the $(-)$ sign associated to each loop, individual configurations may well have a negative weight, so that a straightforward probabilistic interpretation is not available.

Our interest is of course the case when the graph G is the one associated with a discrete domain $(\mathbb{D}, a, b)$ with admissible boundary conditions. If we take for the edge weight K the inverse of the coordination number ν of the tiling, $1 - A$ is essentially the discrete Laplacian with Dirichlet boundary conditions. This suggests again that a continuum limit exists, for which (continuum) loop-erased random walks in a (continuum) domain D from a to b are related to the field-theory of so-called symplectic fermions with measure

$$\mathscr{D}\chi\,\mathscr{D}\bar{\chi}\,\exp\int_{\mathbb{D}}\nabla\chi\cdot\nabla\bar{\chi}$$

with Dirichlet boundary conditions. This field-theory is well-known to be conformally invariant. But $K - 1/\nu = -\lambda^2$ is a scaling function which leads to the addition of a mass term to the action.

We conclude this section by noting without justification that the way to impose the existence of a path from a to b is to insert in correlation functions the observable $J(a)\bar{J}(b)$ where $J(a)$ (resp. $\bar{J}(b)$) is the normal derivative of χ (resp. $\bar{\chi}$) at a (resp. at b).

2.2.4 Another Example of Growth: DLA

Up to now, the two growth processes we have defined shared some common features. The next one, **diffusion-limited aggregation** (DLA), is of a rather different nature. It is believed to have a scale-invariant but not conformally invariant limiting distribution. Another reason to introduce DLA is that it can also be modelled via Loewner chains, a subject we touch in the next section. Finally, DLA seems to be a relevant model for a variety of phenomena in physics, for instance aggregation or deposition phenomena, but also in biology, for instance growth of bacterial colonies under certain circumstances.

DLA stands for *diffusion limited aggregation*. It refers to processes in which the domains grow by aggregating diffusing particles. Namely, one imagines building up a domain by clustering particles one by one. These particles are released from the point at infinity, or uniformly from a large circle around the growing sample, and diffuse as random walkers. They will eventually hit the sample and the first time this happens they stick to it. Then the procedure goes on. By convention, time is incremented by unity each time a particle is added to the domain. Thus at each time step the area of the domain is increased by the physical size of the particle. The position at which the particle is added depends on the probability for a random walker to visit the boundary for the first time at this position, which is essentially what is called the harmonic measure at that position. During this process the clustering domain gets ramified and develops branches and fjords of various scales. The probability for a particle to stick on the cluster is much higher on the tip of the branches than deep inside the fjords. This property, relevant at all scales, is responsible for the fractal structure of the DLA clusters.

In a discrete approach one may imagine that the particles are tiny squares whose centres move on a square lattice whose edge lengths equal that of the particles, so that particles fill the lattice when they are glued together. The centre of a particle moves as a random walker on the square tiling. The probability $Q(x)$ that a particle visits a given tile x satisfies the discrete version of the Laplace equation $\nabla^2 Q = 0$. It vanishes on the boundary of the domain, i.e. $Q = 0$ on the boundary, because the probability for a particle to visit a tile already occupied, i.e. a point of the growing cluster, is zero. The local speed at which the domain is growing is proportional to the probability for a tile next to the interface but on the outer domain to be visited. This probability is proportional to the discrete normal gradient of Q, since the visiting probability vanishes on the interface. So the local speed is $v_n = (\nabla Q)_n$. To add a new particle to the growing domain, a random walk has to wander around and the position at which it finally sticks is influenced by the whole domain. To rephrase this, for each new particle one has to solve the outer Laplace equation, a non-local problem, to know the sticking probability distribution. This is a typical example when scale-invariance is not expected to imply conformal invariance.

It is not so easy to make an unbiased simulation of DLA on the lattice. One of the reasons is that on the lattice there is no such simple boundary as a circle, for which the hitting distribution from infinity is uniform. The hitting distribution on the boundary of a square is not such a simple function. Another reason is that despite the fact that the symmetric random walk is recurrent is $2D$, each walk takes many steps to glue to the growing domain. The typical time to generate a single sample of reasonable size with an acceptable bias is comparable to the time it takes to make enough statistics on loop-erased random walks or percolation to get the scaling exponent with two significant digits. Still this is a modest time, but it is enough to reveal the intricacy of the patterns that are formed. Figure 2.12 is such a sample. The similarity with the sample in Fig. 2.13, obtained by iteration of conformal maps, is striking. But a quantitative comparison of the two models is well out of analytic control and belongs to the realm of extensive simulations. There is now a consensus that the fractal dimension of $2D$ DLA clusters is $d_{f,\text{DLA}} \simeq 1.71$. It has been long debated whether or not discrete simulations done right nevertheless do keep a remnant of the lattice at large distance. There is some consensus now that for instance the orientation of the lattice can be seen even in the large, and rotation invariant algorithms should be preferred.

2.3 Loewner Chains

There are many possible descriptions of subsets of a set. Some may look more natural than others but it is the problem at hand that decides which one is the most efficient. Growth processes in two dimensions involve time-dependent subsets of the complex plane $\mathbb{C}$. Loewner chains have proved to be an invaluable tool in this context. The simplest situation is when they are used to describe families of domains. These notes deal (almost) exclusively with that case.

Loewner chains were introduced (by Loewner!) in the context of the **Bieberbach conjecture**, now a theorem proved by de Branges in 1985. It states that *if $f(z) = z + \sum_{n \geq 2} a_n z^n$ is a holomorphic function injective in the unit disc $\mathbb{U} = \{z \in \mathbb{C}, |z| < 1\}$ then $|a_n| \leq n$ for $n \geq 2$.* Bieberbach proved that $|a_2| \leq 2$ in 1912, and Loewner proved in 1923 that $|a_3| \leq 3$ using a dynamical picture of the changes of $f(\mathbb{U})$ when the a_n's change, starting from the trivial case $f(z) = z$.

Fig. 2.12   A DLA sample

2.3.1 Around Riemann's Theorem

Recall that a domain $\mathbb{D}$ is a non-empty open simply connected strict subset of the complex plane $\mathbb{C}$. Simple connectedness is a notion of purely topological nature which in two dimensions asserts essentially that a domain has no holes and is contractile: a domain has the topology of a disc.

- **Riemann's theorem** states that *two domains $\mathbb{D}$ and $\mathbb{D}'$ are always conformally equivalent*, i.e. there is an invertible holomorphic map $g : \mathbb{D} \mapsto \mathbb{D}'$ between them.

Riemann stated the theorem but his argument had many gaps. This was at least partly at the origin of the formidable development of functional analysis in the twentieth century but it took decades before a proof meeting modern standards was found.

Extending g to the boundary of $\mathbb{D}$ is impossible in general if the topological boundary is used, i.e. if the boundary of $\mathbb{D}$ is taken as the complement of $\mathbb{D}$ in its closure (the notion of boundary one learns at school). As an example, take $\mathbb{D}$ to be the upper half-plane $\mathbb{H}$ with the vertical line segment $]0, ia]$ removed and $\mathbb{D}' = \mathbb{H}$. The naïve boundary of $\mathbb{D}$ is the union of $\mathbb{R}$ and $]0, ia]$. The limits of $g(z)$ when z approaches a given point of the segment $]0, ia]$ from the left or from the right must be distinct. But another notion of boundary can be defined for which a continuous extension at the boundary is always possible. Intuitively, this more involved notion keeps track of the different sides from which a naïve boundary point can be approached.

Fig. 2.13 A shape produced by iteration of random conformal maps

This is trivial in our simple example but the general situation is involved. In Sect. 2.2.1, we mentioned the idea to define an appropriate notion of boundary by duality by saying that the space of functions on the boundary is the space of harmonic functions in the domain. Riemann's theorem gives a strong incentive to do that. It is known that for any continuous function on the unit circle, there is a single function continuous in the closed unit disk, harmonic in the open unit disk and with the prescribed value on the boundary. One can extend this idea to define wilder "functions" on the unit circle by saying that they are "boundary values" of a larger class of harmonic functions in the open unit disk (for instance all bounded harmonic functions). By Riemann's theorem, any domain is conformally equivalent to the open unit disk. As holomorphic maps preserve harmonicity, one gets a notion of "function on the boundary" for an arbitrary domain and this notion is conformally invariant. Then by some kind of duality one passes from the space of functions on the boundary to the boundary itself. Serious work is needed to turn this intuition into mathematics, but it can be done. We shall freely use the word "boundary" in what follows, leaving to the reader the task of deciding from the context which kind of boundary we have in mind. In cases when there is only one way to approach naïve boundary points, the two notions coïncide.

In simple cases, the map f can be found in closed form. For instance, the upper-half-plane $\mathbb{H}$ and the unit disc $\{z \in \mathbb{C}, |z| < 1\}$ centred on the origin are two domains. The conformal transformation $f(z) = i\frac{1-z}{1+z}$ maps the unit disc biholomorphically onto the upper half plane with $f(0) = i$ and $f(1) = 0$. But finding a closed formula for f in the general case is impossible.

The upper half-plane has a three-dimensional Lie group of conformal automorphisms, $PSL_2(\mathbb{R})$, that also acts on the boundary of $\mathbb{H}$. This group is made of homographic transformations $f(z) = \frac{az+b}{cz+d}$ with a, b, c, d real and $ad - bc = 1$. To specify such a map we have to impose three real conditions. Hence, there is a unique holomorphic automorphism—possibly followed by a conjugation—that maps any triple of boundary points to any other triple of boundary points. Similarly there is unique homographic transformation that maps any pair made of a bulk point and a boundary point to another pair of bulk and boundary points. By Riemann's theorem, any domain has a Lie group of conformal automorphisms isomorphic to $PSL_2(\mathbb{R})$ and the same normalisation conditions can be used.

Riemann's theorem is used repeatedly in the rest of these notes. It is the starting point of many approaches to growth phenomena in two dimensions since it allows to code the shapes of growing domains in their uniformising conformal maps. To make the description precise, one has to choose a reference domain against which the growing domains are compared. By Riemann's theorem, we may choose any domain as reference domain—and depending on the geometry of the problem some choices are more convenient than others. The unit disc and the upper half plane are often used as reference domains.

2.3.2 Hulls

One can be a more explicit when the domain $\mathbb{D}$ differs only locally from the upper half-plane $\mathbb{H}$, that is if $\mathbb{K} = \mathbb{H} \setminus \mathbb{D}$ is bounded. Such a set $\mathbb{K}$ is called a **hull**. The real points in the closure of $\mathbb{K}$ in $\mathbb{C}$ form a compact set which we call $\mathbb{K}_\mathbb{R}$. In that case, $\mathbb{H}$ is the convenient reference domain. Let $g : \mathbb{D} \mapsto \mathbb{H}$ be a conformal bijection. For $z \in \mathbb{D}$ define $g(\bar{z}) := \overline{g(z)}$. If z approaches a point x on the real axis while staying within $\mathbb{D}$, $g(z)$ has a real limit which we denote by $g(x)$. It follows that g extends to a holomorphic map on the connected open set $\mathbb{D} \cup \overline{\mathbb{D}} \cup (\mathbb{R} \setminus \mathbb{K}_\mathbb{R}) \cup \infty$ of the Riemann sphere, which contains a neighbourhood of ∞. This is an illustration of the Schwartz reflection principle. One can use the automorphism group of $\mathbb{H}$ to ensure that $g(z) = z + O(1/z)$ for large z. This is called the **hydrodynamic normalisation**. It involves three conditions: g maps ∞ to ∞, has unit derivative there, and has no constant term. These three condition are real because ∞ is on the boundary of the upper half-plane seen within the Riemann sphere. There is no further freedom left. Thus any property of g is an intrinsic property of $\mathbb{K}$.

We shall denote this special representative by $g_\mathbb{K}$. The inverse map $f_\mathbb{K}$ is holomorphic on the full Riemann sphere except for cut along a compact subset of $\mathbb{R}$ across which its imaginary part has a positive discontinuity (in general this is a measure) $d\mu(x)$. Away from the cut, $f_\mathbb{K}$ has the standard representation

$$f_\mathbb{K}(w) = w - \frac{1}{\pi} \int_\mathbb{R} \frac{d\mu(x)}{w - x}.$$

The coefficients of the expansion of $f_\mathbb{K}$ at infinity are essentially the moments of μ. I particular, they are real. Each of them quantifies an intrinsic property of $\mathbb{K}$. The number $C_\mathbb{K} := \frac{1}{\pi} \int_\mathbb{R} d\mu(x)$ is the **total mass** of μ. It is positive (or 0 is $\mathbb{K}$ is empty). Note that $f_\mathbb{K}(w) = w - C_\mathbb{K}/w + \ldots$ at large w and by inverting, $g_\mathbb{K}(z) = z + C_\mathbb{K}/z + \ldots$ at large z. The coefficient $C_\mathbb{K}$ plays an important role. It is called the **capacity** of $\mathbb{K}$ seen from ∞. It's positivity is intuitively related to the fact that one *removes* a piece from $\mathbb{H}$. Capacity is trivially translation-invariant, $(C_{x+\mathbb{K}} = C_\mathbb{K}$ where, for $x \in \mathbb{R}$, $x + \mathbb{K}$ denotes the translate of $\mathbb{K}$ by x units along the real axis) and has weight 2 under dilatations $(C_{s\mathbb{K}} = s^2 C_\mathbb{K}$ if s is a positive scale factor). Capacity has an additive property: simple series manipulations show that if $\mathbb{K}'$ and $\mathbb{K}''$ are two hulls and $\mathbb{K} = \mathbb{K}' \cup g_{\mathbb{K}'}^{-1}(\mathbb{K}'')$ (which is another hull) then $C_\mathbb{K} = C_{\mathbb{K}'} + C_{\mathbb{K}''}$.

2.3.3 Basic Examples

Example 1 (The semi-disc) Maybe the simplest example is when $\mathbb{K}$ is a semi-disc $\{z \in \mathbb{H}, |z - b| \leq r\}$ for a real b and real positive r. Then $g_\mathbb{K}(z) = z + r^2/(z - b)$. Expansion at large z shows that $C_\mathbb{K} = r^2$.

Example 2 (The vertical line segment) In the example when $\mathbb{K}$ is the vertical line segment $]0, ia]$, one gets $g_\mathbb{K}(z) = \sqrt{z^2 + a^2}$, a formula by which we mean the analytic continuation of the function $z\sqrt{1 + a^2/z^2}$ were the square root is defined by its usual power series around 1 when z is large. Expansion at large z shows that $2C_\mathbb{K} = a^2$.

Example 3 (The oblique line segment) The case when $\mathbb{K}$ is an oblique line segment $]0, ae^{i\pi b}]$ making an angle πb with respect to the real positive axis ($b \in \,]0, 1[$) yields

$$z = \left(g_\mathbb{K}(z) - x_+\right)^b \left(g_\mathbb{K}(z) - x_-\right)^{1-b},$$

where the real parameters $x_- < 0 < x_+$ satisfy $bx_+ + (1 - b)x_- = 0$ and $(-x_-)^b x_+^{1-b} = a$. Expansion at large z shows that $2C_\mathbb{K} = b(1 - b)(x_+ - x_-)^2$. The closer the line is to the real axis (i.e. the closer b is to 0 or π) and the larger a has to be to reach a given capacity.

Example 4 (Arc of circle) An instructive example is when $\mathbb{K}$ is the arc $]r, re^{i\vartheta}]$ of a circle centred at 0 of radius r. Some of the following computations require to keep a precise track of the determination of the square root that appears in the formula for $g_\mathbb{K}$ because it is crucial for the interpretation. The map $f(w) = (w - r)/(w + r)$ sends the arc to the vertical line segment $]0, i \tan \vartheta/2]$, so that by Example 2, $w \mapsto \sqrt{f(w)^2 + \tan^2 \vartheta/2}$ is a conformal map from $\mathbb{D}$ to $\mathbb{H}$. However, this map sends ∞ to $1/(\cos \vartheta/2)$, not to ∞. To get the hydrodynamic normalisation, we have to compose with an appropriate automorphism of $\mathbb{H}$. This yields

$$g_\mathbb{K}(w) = r \frac{-(2 - \cos^2 \vartheta/2) \cos \vartheta/2 \sqrt{(\frac{z-r}{z+r})^2 + \tan^2 \vartheta/2} + 2 - 3\cos^2 \vartheta/2}{\cos \vartheta/2 \sqrt{(\frac{z-r}{z+r})^2 + \tan^2 \vartheta/2} - 1},$$

whose expansion at ∞ starts like $g_\mathbb{K}(w) = w + (1 - \cos^4 \vartheta/2)r^2/w + O(1/w^2)$. Hence the capacity is $C_\mathbb{K} = (1 - \cos^4 \vartheta/2)r^2$.

The tip of the arc, $re^{i\vartheta}$ is mapped to $(3\cos^2\vartheta/2 - 2)r$ by $g_{\mathbb{K}}$. One checks that

$$\left(g_{\mathbb{K}}(w) - g_{\mathbb{K}}\left(re^{i\vartheta}\right)\right)\frac{\partial g_{\mathbb{K}}(w)}{\partial\vartheta} = 2r^2\sin\vartheta/2\cos^3\vartheta/2,$$

which is w-independent.

Moreover $\lim_{w\mapsto r^-} g_{\mathbb{K}}(w) = r(1 - 2\sin\vartheta/2 - \sin^2\vartheta/2)$ and $\lim_{w\mapsto r^+} g_{\mathbb{K}}(w) = r(1 + 2\sin\vartheta/2 - \sin^2\vartheta/2)$. The behaviour of $g_{\mathbb{K}}$ when $\vartheta\mapsto\pi^-$ is interesting. In this limit, $\mathbb{K}$ becomes a semicircle. Let $\tilde{\mathbb{K}} = \{w\in\mathbb{H}, |w|\le r\}$ be the corresponding semi-disc. The points w inside $\tilde{\mathbb{K}}$ are cut away from ∞ when $\vartheta\mapsto\pi^-$, and one checks that $\lim_{\vartheta\mapsto\pi^-} g_{\mathbb{K}}(w) = -2r$ for these points, i.e. they are swallowed in the limit. However, the points $\{w\in\mathbb{H}, |w|>r\}$ are mapped to $\lim_{\vartheta\mapsto\pi^-} g_{\mathbb{K}}(w) = w + r^2/w = g_{\tilde{\mathbb{K}}}(w)$.

2.3.4 Iteration of Conformal Maps

With Riemann's theorem at our disposal, we can start to encode growth processes. Suppose than the initial domain is the upper half-plane and that a small amount of matter is removed at each time step (so that in fact it is the lower half-plane that grows). At time step n, a certain $\mathbb{K}_n$ has been removed from $\mathbb{H}$. Let $g_n := g_{\mathbb{K}_n}$ denote the corresponding map and f_n its inverse. Then $g_n(\mathbb{K}_{n+1}\setminus\mathbb{K}_n)$ describes a small amount of matter removed to $\mathbb{H}$. If $g_n(\mathbb{K}_{n+1}\setminus\mathbb{K}_n)$ has typical size s and is located in the neighbourhood of point x on the real axis, $\mathbb{K}_{n+1}\setminus\mathbb{K}_n$, which is what is really removed at time $n+1$ has typical size $s|f_n'(x)|$.

Example 5 (Simple iteration) Choose a small number ε. Let b_n, $n>0$ be an independent sequence drawn from some chosen probability distribution. At time step $n+1$ take $g_n(\mathbb{K}_{n+1}\setminus\mathbb{K}_n)$ to be the semi-disc $\{z\in\mathbb{H}, |z - b_{n+1}||f_n'(b_{n+1})|\le\varepsilon\}$, so that

$$g_{n+1}(z) = g_n(z) + \frac{\varepsilon^2}{|f_n'(b_{n+1})|^2(g_n(z) - b_{n+1})}.$$

This defines a random growth process were at each time step a small semi-disc-like grain of matter of size $\sim\varepsilon$ is removed. Despite its simplicity, little is known (at least to the author) about this process.

Many other (probabilistic or deterministic) rules can be invented, but the resulting processes are mostly impossible to study analytically at the moment. Let us simply note to conclude that the samples obtained by methods (but using the disc geometry) look strikingly like DLA. Figure 2.13 is obtained by iteration of conformal maps, compare with Fig. 2.12.

2.3.5 Continuous Time Growth Processes

Our aim is to motivate the introduction of Loewner chains.

Fig. 2.14 The uniformisation of well-separated small semi-discs

If $\mathbb{K}$ is not simply a semi-disc, but a union of well-separated small semi-discs of radii r_α centred at b_α (see Fig. 2.14), a moment of thought leads to realise that

$$g_\mathbb{K}(z) \sim z + \sum_\alpha \frac{r_\alpha^2}{z - b_\alpha}.$$

The large z expansion yields $C_\mathbb{K} \sim \sum_\alpha r_\alpha^2$, a positive number as expected.

Taking a naïve limit, one gets that if ε is a small positive number, $v(x)$ is a non-negative function on $\mathbb{R}$ and $\mathbb{K} = \{z = x + iy \in \mathbb{H}, \ y \leq \varepsilon v(x)\}$ then

$$g_\mathbb{K}(z) \sim z + \frac{\varepsilon}{\pi} \int_\mathbb{R} \frac{v(u)\,du}{z - u}.$$

Indeed, using that, if $v(x) \neq 0$, $\lim_{\varepsilon \to 0^+} \operatorname{Im} m(x + i\varepsilon v(x) - u)^{-1} = \pi\delta(u - x)$ one checks that $\operatorname{Im} m \frac{1}{\pi} \int_\mathbb{R} \frac{v(u)\,du}{x + i\varepsilon v(x) - u} \sim -v(x)$ so that to first order in ε $g_\mathbb{K}(z)$ is real when z is on the boundary of $\mathbb{K}$. Even more generally, one could replace the positive measure $v(u)\,du$ by any positive measure $d\rho(u)$. A naïve large-z expansion, certainly valid if the function v (or more generally the measure $d\rho$) has compact support and finite mass, gives $C_\mathbb{K} \sim \frac{\varepsilon}{\pi} \int_\mathbb{R} v(u)\,du$ (more generally $C_\mathbb{K} \sim \frac{\varepsilon}{\pi}\rho(\mathbb{R})$), again a positive number.

Now think about a continuous time growth process for which $\mathbb{K}_t$ has been removed from $\mathbb{H}$ at time t. Set $\mathbb{H}_t := \mathbb{H} \setminus \mathbb{K}_t$. Let $g_t := g_{\mathbb{K}_t} : \mathbb{H}_t \to \mathbb{H}$ denote the corresponding map and $f_t : \mathbb{H} \to \mathbb{H}_t$ its inverse. Fix t and a small positive ε. Then $g_t(\mathbb{K}_{t+\varepsilon} \setminus \mathbb{K}_t)$ describes a small amount of matter removed to $\mathbb{H}$. We could take as a definition of continuous time growth that the associated map $g_{t+\varepsilon} \circ f_t$ is described by a non-negative function $v_t(u)$ or more generally a positive measure $d\rho_t(u)$ as above. Taking the limit $\varepsilon \mapsto 0^+$ leads to

$$\frac{\partial g_t(z)}{\partial t} = \frac{1}{\pi} \int_\mathbb{R} \frac{d\rho_t(u)}{g_t(z) - u}. \tag{2.7}$$

Taking the time derivative of $f_t \circ g_t(z) = z$ and substituting $w = g_t(z)$ yields

$$\frac{\partial f_t(w)}{\partial t} = -f_t'(w)p(w, t) \quad \text{where } p(w, t) := \frac{1}{\pi} \int_\mathbb{R} \frac{d\rho_t(u)}{w - u}. \tag{2.8}$$

Note that $p(w, t)$ is holomorphic in $\mathbb{H}$ and the positive measure $\rho_t(u)$ is its boundary value in a generalised sense (hyperfunctions).

Equation (2.8) is called a **Loewner chain with reference domain** $\mathbb{H}$, though we shall use the name *Loewner chain* for (2.7) as well. The analogous equations with reference domain the unit disc can be obtained straightforwardly by the same arguments. The large-z expansion yields

$$\frac{dC_{\mathbb{K}_t}}{dt} = \frac{1}{\pi}\rho_t(\mathbb{R}) \geq 0.$$

So if hulls are constructed little by little by a growth process, the capacity increases with time (in particular it is obviously positive).

In principle, if the family of measures ρ_t is given, one can solve for $g_t(z)$ with the initial condition $g_0(z) = z$. Again, ρ_t can be random or deterministic. We should note that Loewner chains are in some sense kinematic equations that give a general framework to encode growth processes. But in a real dynamical problem ρ_t has to be specified. It may depend explicitly on g_t. For instance $d\rho_t(u) = |f'_t(u)|^{-2} du$ is related to Laplacian growth, though the unit disc geometry is the relevant one in that case. The exponent -2, which we already interpreted for discrete iteration, ensures that the size of $\mathbb{K}_t$ grows linearly with time. But other exponents between 0 and -2 are interesting too. Note that DLA provides a discrete analogue of Laplacian growth. The particle size plays the role of an ultraviolet cutoff.

2.3.6 Geometric Interpretation

One can give the following geometric interpretation of Loewner chains. Set $g_t(z) := z_t$, view z_t as the position of a fluid particle as time goes by, and suppose for simplicity that $d\rho_t(u) = v_t(u) du$ so that the Loewner chain becomes

$$\frac{dz_t}{dt} = \frac{1}{\pi} \int_{\mathbb{R}} \frac{v_t(u) du}{z_t - u}. \tag{2.9}$$

Hence $\frac{1}{\pi} \int_{\mathbb{R}} \frac{v_t(u) du}{z - u}$ plays the role of a time-dependent holomorphic vector field on the manifold with boundary $\mathbb{H}$. At point $z = x + i0^+$ i.e. close to the real axis (the boundary of $\mathbb{H}$) this vector field has imaginary part $-v(x)$, so that when x is away from the support of ρ_t, (that is, when $v_t(.) = 0$ in a neighbourhood of x), the vector field is real, i.e. tangent to the boundary. However, if x is on the support of ρ_t the vector field has a finite negative imaginary part, which means that some fluid particles that started inside $\mathbb{H}$ can be swallowed by the boundary. In fact $\mathbb{K}_t$ is nothing but the set of fluid particles which where in $\mathbb{H}$ at $t = 0$ but have hit the boundary before time t.

The reader is urged to review Examples 1–4 in this light. For the semi-disc case, take r as time, either with $b = 0$ or with $b = r$. For the case of line segments, take a constant b and use a as time. For the arc of circle, using ϑ as time, with special care in the limit $\vartheta \mapsto \pi^-$. It is instructive to compute the measure ρ_t in each case and to check that the above interpretation of $\mathbb{K}_t$ is correct.

Another, more abstract, geometric interpretation is also possible. Let N_- be the group of series of the form $z + \sum_{m \leq -1} g_m z^{m+1}$ with real coefficients and convergent for large z (the domain of convergence may depend on the series, so N_- is made of "germs", and is in fact the group of germs of holomorphic functions fixing ∞ and with derivative 1 at ∞). In the same spirit, let O_∞ be the space of germs of holomorphic functions at infinity. We let N_- act on O_∞ by composition, $\gamma_g \cdot F := F \circ g$. Observe that $\gamma_{g_1 \circ g_2} = \gamma_{g_2} \cdot \gamma_{g_1}$ so this is an anti-representation.

Note that the g_t's of a Loewner chain with bounded $\mathbb{K}_t$ belong to N_-. If $F \in O_\infty$ and if z is large enough, $F(z)$ is well defined as well as $F(z_t)$ for small t (where the meaning of small may depend on z and F) and

$$\frac{dF(z_t)}{dt} = \frac{1}{\pi} \int_{\mathbb{R}} \frac{d\rho_t(u)}{z_t - u} \frac{\partial F}{\partial z}(z_t),$$

which can be rewritten

$$\frac{d}{dt}(\gamma_{g_t} \cdot F) = \gamma_{g_t} \cdot (v_t \cdot F)$$

where $v_t(z) := \frac{1}{\pi} \int_{\mathbb{R}} \frac{d\rho_t(u)}{z-u} \frac{\partial}{\partial z}$ is a germ of vector field.

So the Loewner chain equation can be viewed as a flow on N_-

$$\frac{d}{dt} \gamma_{g_t} = \gamma_{g_t} \cdot v_t.$$

The group N_- has an interesting representation theory, related to that of the Virasoro algebra, which can be used as a probe for this flow.

2.3.7 Local Growth

Suppose that as time goes by the measures ρ_s are δ-peaks of height $2\pi a_s$ (the factor 2 is purely historical) at position ξ_s: in physicist notation $d\rho_s(u) = 2\pi a_s \delta(u - \xi_s) du$. In the upper half plane reference geometry, the growth process will be described by an equation of the type

$$\frac{\partial g_s(z)}{\partial s} = \frac{2a_s}{g_s(z) - \xi_s}. \tag{2.10}$$

Note that Examples 2–4 fall in this category. The formula was given for Example 4 if $s = \vartheta$ and the other cases lead to simple computations left to the reader.

If one is interested only in the growth of the hull, but not in the way the evolution is parameterised, one can make change the time variable without arm. The statement that ξ_s changes quickly or slowly makes sense only compared with the changes in a_s. For instance, suppose that the function a_s vanishes in some interval, while ξ_s keeps on changing so that it has a different value at the beginning and at the end of the interval. During that interval g_s has not changed but when a_s starts moving again, the place at which the hull resumes growth can be far from the place where it was growing before the pause. This is a limiting case of what happens when variations of ξ_s are large with respect to those of a_s. This means that if, at s_0, ξ_s starts to move very fast with respect to a_s, the growth takes place very near $\mathbb{K}_{s_0}$ or the real axis. This conclusion is supported by Example 3.

We also infer that to have local growth, i.e. to have the position where the hull grows vary continuously, we need to impose that ξ_s stops if a_s does. To make this statement precise, it is convenient to go to a special time parameterisation. The capacity of the hull at time s is $C_{\mathbb{K}_s} = 2 \int_0^s ds' \, a_{s'}$, a non-decreasing function of s.

Define $t = \int_0^s \mathrm{d}s' \, a_{s'}$, take t to be the new time variable and by abuse of notation write ξ_t for $\xi_{s(t)}$, $\mathbb{K}_t$ for $\mathbb{K}_{s(t)}$ and so on. Then by construction $C_{\mathbb{K}_t} = 2t$ and the equation reads

$$\frac{\partial g_t(z)}{\partial t} = \frac{2}{g_t(z) - \xi_t}. \tag{2.11}$$

We take as a definition of **local growth** that ξ_t is continuous function of t. The function ξ_t is often called the **driving function** of the Loewner evolution. It is sometimes convenient to normalise ξ_t by $\xi_0 = 0$ or what amounts to the same to impose that the hull starts growing from point 0.

A broad class of growing hulls that can be described by such an equation is given by continuous simple curves started on the boundary of $\mathbb{H}$ and staying in $\mathbb{H}$ thereafter. Let $\gamma_{]0,\infty]}$ be a parameterised simple continuous curve from 0 to ∞ in $\mathbb{H}$ and assume that the capacity parameterisations has been chosen, so that $\mathbb{K}_t := \gamma_{]0,t]}$ is a hull with capacity $2t$. When ε is small, $\mathbb{K}_{\varepsilon,t} := g_t(\gamma_{]t,t+\varepsilon]})$ is a tiny piece of a curve. The support of the discontinuity measure $\mathrm{d}\rho_{f_{\varepsilon,t}}$ is small and becomes a point when ε goes to 0. Measures supported at a point are δ functions, so there is a point ξ_t such that, as a measure, $\mathrm{d}\rho_{f_{\varepsilon,t}}/\mathrm{d}x \sim 2\varepsilon\delta(x - \xi_t)$ as $\varepsilon \to 0^+$.

For a general local Loewner growth process, one defines $\gamma_t = f_t(\xi_t + i0^+) := \lim_{\varepsilon \mapsto 0^+} f_t(\xi_t + i\varepsilon)$ (remember f_t is the inverse map of g_t). We shall often use the shorthand notation $\gamma_t = f_t(\xi_t)$. The set $\gamma_{]0,t]} := \bigcup_{s \in]0,t]} \gamma_s$ is called the **trace** of the growth process. If the hull is a simple curve, the notation is consistent. Whether the trace is a curve (simple or not) in general is highly non-obvious, but this will be the case for all examples in these notes, though proving it can be a formidable task.

At time t, growth takes place at point ξ_t in the g_t plane i.e. at point γ_t in the original "physical" plane. Thus it is tempting to conclude that $\mathbb{K}_t$ coincides with $\gamma_{]0,t]}$. Though this picture works nicely for Examples 2–3, it is slightly too naïve and fails in Example 4 when the trace, which is an arc of circle closes to a semicircle and the corresponding semi-disc completes the hull.

For a given z with $\mathrm{Im}\, mz \geq 0$ and $z \neq \xi_0$, the local existence and uniqueness of solutions to Eq. (2.11) is granted by general theorems on ordinary differential equations, but problems may arise if a time τ_z (depending on z in general) exists for which $g_{\tau_z}(z) = \xi_{\tau_z}$. One possibility is to declare $g_t(z)$ undefined for $t \geq \tau_z$. But it is often the case that, as suggested by Examples 2–3, the two limits $\lim_{x \mapsto \xi_{\tau_z}^\pm} g_t \circ f_{\tau_z}(x)$ exist, allowing to think that after τ_z, $g_t(z)$ has split in two real trajectories.

There is a **regularity criterion** on the function ξ that guaranties that if $x \neq \xi_0$ is real, τ_x is infinite. It is sufficient that for each t,

$$\lim_{s \mapsto t^-} \sup_{t' \in [s,t[} \frac{|\xi_t - \xi_{t'}|}{|t - t'|^{1/2}} < 4. \tag{2.12}$$

To prove this criterion, it is convenient to consider $X_t := g_t(x) - \xi_t$, a continuous function which satisfies the integral equation $X_t = x - \xi_t + \int_0^t \frac{2\,\mathrm{d}s}{X_s}$. As this implies that $\xi_\tau - \xi_t = X_t - X_\tau + \int_t^\tau \frac{2\,\mathrm{d}s}{X_s}$, we can see ξ as a functional of X. The task is to control its behaviour if X_t has a given sign, say positive, on $[0, \tau[$ and vanishes at τ. It is clear that the two terms in $X_t + \int_t^\tau \frac{2\,\mathrm{d}s}{X_s}$ vary in opposite directions, in that

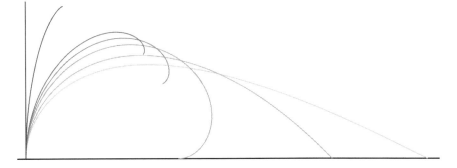

Fig. 2.15 The hull at time τ for $\alpha = 0, 1, 2, 3, 4, 5, 6$

the faster X_t goes to 0, the slower is the vanishing of $\int_t^\tau \frac{2\,ds}{X_s}$ at $t = \tau$. So the mildest behaviour of the sum as t goes to τ is when the two terms have a similar behaviour. A detailed analysis requires some care, but a quick and dirty way to retrieve the criterion is to impose that the two terms be equal, which gives $X_t = 2\sqrt{\tau - t}$ hence $\xi_\tau - \xi_t = 4\sqrt{\tau - t}$ as announced.

Example 6 (Square root driving term) The Loewner equation when $\xi_\tau - \xi_t = 4\alpha\sqrt{\tau - t}$ can be solved in closed form for any α though the formulæ are cumbersome. We normalise ξ_t so that $\xi_0 = 0$, i.e. take $\xi_t = 4\alpha(\sqrt{\tau} - \sqrt{\tau - t})$. By left-right symmetry, we can assume that $\alpha \geq 0$. For $\alpha \in [1, +\infty[$ it is convenient to set $\alpha := \cosh\eta, \eta \in [0, +\infty[$. One parameterises time as

$$\frac{2e^{-\eta\coth\eta}\sinh\eta}{\sin(2\vartheta\sinh\eta)}\frac{(\sin(\vartheta e^\eta))^{(\coth\eta+1)/2}}{(\sin(\vartheta e^{-\eta}))^{(\coth\eta-1)/2}} = \sqrt{\frac{\tau - t}{\tau}},$$

with $\vartheta \in [0, \pi e^{-\eta}]$. As a function of ϑ, the hull builds the curve

$$\left\{2\sqrt{\tau}\left(e^{-\eta} - \frac{2\sinh\eta\sin(\vartheta e^{-\eta})}{\sin(2\vartheta\sinh\eta)}e^{i\vartheta e^\eta}\right)\right\}_{\vartheta\in[0,\pi e^{-\eta}]}.$$

For $\vartheta = \pi e^{-\eta}$ the curve closes a whole domain, just as in the arc of circle Example 4, which in fact is the special case $\alpha = 3\sqrt{2}$.

For $\alpha \in [0, 1[$ it is convenient to set $\alpha := \cos\varphi, \varphi \in]0, \pi/2]$. The formulæ can be obtained by analytic continuation $\eta \mapsto i\varphi$, this time with a parameter $\vartheta \in [0, \infty]$. The hulls remain simple curves even for $\vartheta = \infty$.

Figure 2.15 illustrates the different behaviours.

The very same criterion on the behaviour of the function $\xi_.$ is also sufficient to ensure that the hull $\mathbb{K}_t$ is a simple continuous curve, say $\{\gamma_s, s \in]0, t]\}$, and $\gamma_t = f_t(\xi_t)$, i.e. that our naïve expectation $\mathbb{K}_t = \bigcup_{s\in]0,t]} f_s(\xi_s)$ is fulfilled.

The two properties—"$g_t(x)$ for real x does not hit ξ_t" and "the hull is a simple curve"—are in fact equivalent. The intuitive reason is the following. The fact that $g_t(x)$ for real x hits ξ_t at some time τ is the sign that at time τ the hull "swallows a whole piece of $\mathbb{H}$". The previous example illustrates this relationship when the hull hits the real axis. But from the point of view of iteration, if $s \geq 0$ is fixed, it is

Fig. 2.16 A *curve* hitting itself or the *real axis* a number of times

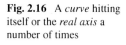
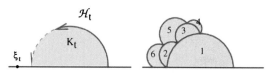

obvious that when $t \geq 0$ varies the function $\tilde{g}_{t,s}(z) := g_{t+s} \circ f_s(z + \xi_s) - \xi_s$ satisfies the Loewner equation (2.11) with driving function $\tilde{\xi}_t := \xi_{t+s} - \xi_s$. So if the driving function $\tilde{\xi}_t := \xi_{t+s} - \xi_s$ leads to a hull hitting the real axis, the driving function ξ_t leads to a hull hitting itself or the real axis, as illustrated in Fig. 2.16. This discussion also explains why, if the trace is a continuous curve, it can have double points but no crossings.

2.4 Stochastic Loewner Evolutions

Stochastic Loewner evolutions were introduced by Schramm in 1999 as a general framework to study random curves satisfying certain properties. His specific interest was to prove that loop-erased random walks on a two-dimensional lattice have a conformally invariant continuum limit. Schramm observed that these walks have on the lattice the so-called *domain Markov property* (to be defined below) a property that can be rephrased in the continuum. Though he was not able at that time to prove the existence of a conformally invariant limit of loop-erased random walks, he recognised that conformal invariance and the domain Markov property brought together would have remarkable consequences, and was able to prove that the probability measures on random curves in the continuum satisfying at the same time conformal invariance and the domain Markov property formed a one-parameter family. Crucial to the proof and the explicit description of these measures was the idea of viewing curves as hulls and to use Loewner evolutions. That in this context the most useful description of a curve is by encoding it into a growth process via a Loewner chain is at first sight very surprising and may explain why physicists who had understood the importance of conformal invariance to study many examples of random curves in the early 1980's failed to "produce Schramm's argument before Schramm".

The general idea is to impose properties relating different members in a family of probability measures on continuous curves without crossings, but possibly with multiple points. Let us note that curves here are considered modulo reparameterisations, but not simply as subsets of the plane. For simple curves, this would essentially make no difference, but curves with multiple points require more care.

In the discrete setting, it is a fact that interfaces on appropriate lattices are simple curves, so why bother to deal with non-simple curves? The answer is that even if at the scale of the lattice spacing the interface is simple, when one tries to take a continuum limit by looking at a macroscopic scale while taking a smaller and smaller lattice spacing, a curve that makes a large excursion and then comes back close to itself, say a few lattice spacings away, has a double point from the macroscopic

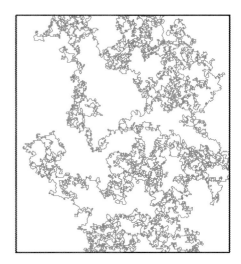

viewpoint. While in some models—like loop-erased random walks, Schramm's initial motivation—the interface remain simple when the lattice spacing gets smaller, some other models—like percolation—clearly exhibit multiple points in the continuum limit. This is clearly seen on samples, see Fig. 2.17.

In the following three sections, we suppose that we are given a family of probability measures $\{\mathbf{P}_{\mathbb{D},a,b}\}$ indexed by triples consisting of a domain $\mathbb{D}$ and two distinct boundary points a, b of $\mathbb{D}$. For a given triple $(\mathbb{D}, a, b)$, $\mathbf{P}_{\mathbb{D},a,b}$ is a measure on $\Omega_{\mathbb{D},a,b}$, the set of continuous curves without crossings within $\overline{\mathbb{D}}$—the union of $\mathbb{D}$ and its boundary (in the refined sense alluded too in Sect. 2.3.1)—joining a to b (it is understood that a and b are not multiple points).

First, we want do define what it means for the family $\{\mathbf{P}_{\mathbb{D},a,b}\}$ to be conformally invariant and to have the domain Markov property.

2.4.1 Conformal Invariance

By Riemann's theorem, if $(\mathbb{D}, a, b)$ and $(\mathbb{D}, a', b')$ are any two triples, there is a conformal map $g : \mathbb{D} \mapsto \mathbb{D}'$ such that $g(a) = a'$ and $g(b) = b'$. It is clear that g induces a bijection, which we call $\breve{g}$, from $\Omega_{\mathbb{D},a,b}$ to $\Omega_{\mathbb{D}',a',b'}$. **Conformal invariance** of the family $\{\mathbf{P}_{\mathbb{D},a,b}\}$ is the statement that $\breve{g}$ is measurable and the image measure $\mathbf{P}_{\mathbb{D},a,b} \circ \breve{g}^{-1}$ coincides with $\mathbf{P}_{\mathbb{D}',a',b'}$, i.e. if C' is a measurable subset of $\Omega_{\mathbb{D}',a',b'}$ then $\breve{g}^{-1}(C')$ is a measurable subset of $\Omega_{\mathbb{D},a,b}$ and $\mathbf{P}_{\mathbb{D},a,b}(\breve{g}^{-1}(C')) = \mathbf{P}_{\mathbb{D}',a',b'}(C')$.

Conformal invariance by itself is a rather weak constraint. Indeed, suppose that a probability $\mathbf{P}_{\mathbb{D}_0,a_0,b_0}$ on $\Omega_{\mathbb{D}_0,a_0,b_0}$ has been defined for a single triple $\mathbb{D}_0, a_0, b_0$ and that it is invariant under the conformal transformations of $\mathbb{D}_0$ fixing a_0 and b_0. Such transformations form a group with one real parameter. Then the direct image $\mathbf{P}_{\mathbb{D}_0,a_0,b_0}$ by any conformal transformation g will define unambiguously

$\mathbf{P}_{g(\mathbb{D}_0),g(a_0),g(b_0)}$. By the Riemann mapping theorem, this defines $\mathbf{P}_{\mathbb{D},a,b}$ for any triple, and the resulting family of probabilities is clearly conformally invariant.

To get a more rigid situation, one has to impose another constraint on the family $\{\mathbf{P}_{\mathbb{D},a,b}\}$. Schramm translated in the continuum a property that holds for loop-erased random walks in the discrete setting: the **domain Markov property**, to which we turn our attention now.

Before doing so, let us remark that this strategy is rather typical. If continuous curves without crossings are replaced by general hulls joining a to b in $\mathbb{D}$ the notion of domain Markov property does not make sense but another one, restriction, turns out to be fruitful and allow for another complete classification. We shall have little to say about these nice "restriction measures" in these notes.

2.4.2 Domain Markov Property

Fix a triple $(\mathbb{D}, a, b)$ and consider an element $\gamma \in \Omega_{\mathbb{D},a,b}$. If a real continuous parameter along γ is given and s is any intermediate value of the parameter, the past and the future of s split γ in two (not necessarily disjoint) curves without crossings. The curve corresponding to the past of s starts at a and is called an **initial segment** of γ. The curve corresponding to the future of s ends at b and is called a **final segment** of γ. The final segment starts at some point $c \in \mathbb{D}$ which is also the end of the initial segment. We use the notation $\gamma_{|a,c|}$ for such an initial segment with point c included and $\gamma_{|c,b|}$ for the final segment. Beware that the notation is a bit ambiguous, because of possible multiple points on γ.

Several curves γ' share the same initial segment $\gamma_{|a,c|}$, and the discussion that follows focuses on the question: if an initial segment is given, what is the distribution of the final segment?

Making sense of this question is not so obvious. First, there should be enough measurable sets in $\Omega_{\mathbb{D},a,b}$. We shall for a while assume that this is so. But even in that case, the event "γ' starts exactly with $\gamma_{|a,c|}$" is more than likely to occur with probability 0. Vaguely, what may have a non-trivial probability is the event "γ' has an initial segment that is close (in some quantified sense) to $\gamma_{|a,c|}$". Probabilists have invented so-called *conditional expectations* and regular conditional probabilities just to deal with that kind of situations. Starting from $\mathbf{P}_{\mathbb{D},a,b}$ this allows to define new probability measures, denoted $\mathbf{P}_{\mathbb{D},a,b}(\,|\gamma_{|a,c|})$, read "conditional probability given the initial segment $\gamma_{|a,c|}$", that can be manipulated just as conditional probabilities when the state space is discrete.[3]

The set of points in $\mathbb{D}$ that cannot be joined to b by a continuous curve in $\mathbb{D}$ without hitting the initial segment form a set that we call a **hull**[4] and denote by $\mathbb{K}_c$.

---

[3] There is a small price to pay, however. For instance, the definition of this conditional probability may fail or be ambiguous for certain $\gamma_{|a,c|}$ but these nasty initial segments form altogether a set of probability 0 for $\mathbf{P}_{\mathbb{D},a,b}$.

[4] If $(\mathbb{D}, a, b) = (\mathbb{H}, 0, \infty)$, this is consistent with our initial definition, and with the new definition, conformal maps send hulls to hulls.

This notation is again slightly ambiguous. Note that $\mathbb{D} \setminus \mathbb{K}_c$ is again a domain. If the initial segment is $\gamma_{|a,c|}$, the final segment starts at c and never enters inside $\mathbb{K}_c$. So the support of the conditional probability $\mathbf{P}_{\mathbb{D},a,b}(\cdot | \gamma_{|a,c|})$ is included in $\Omega_{\mathbb{D} \setminus \mathbb{K}_c,c,b}$. But on this set we have another probability measure, namely $\mathbf{P}_{\mathbb{D} \setminus \mathbb{K}_c,c,b}$, and the two can be compared.

We say that a set $\{\gamma_{|a,c|}\}$ of curves in $\mathbb{D}$ without crossings starting at a is a set of distinct representatives if any curve in $\Omega_{\mathbb{D},a,b}$ has exactly one of its initial segments in $\{\gamma_{|a,c|}\}$. For instance, for the triple $(\mathbb{H}, 0, \infty)$, the initial segments whose associated hull has capacity t form a set of distinct representatives. Intuitively, to get the expectation of a random variable on $\Omega_{\mathbb{D},a,b}$, one can compute its conditional expectation on $\gamma_{|a,c|}$, and then integrate over $\gamma_{|a,c|}$ in a system of distinct representatives.

The family $\{\mathbf{P}_{\mathbb{D},a,b}\}$ is said to have the **domain Markov property** if, for any triple $(\mathbb{D}, a, b)$, one has

$$\mathbf{P}_{\mathbb{D},a,b}(\cdot | \gamma_{|a,c|}) = \mathbf{P}_{\mathbb{D} \setminus \mathbb{K}_c,c,b}, \qquad (2.13)$$

except maybe for a set of initial segments whose intersection with any system of distinct representatives is of measure 0 for $\mathbf{P}_{\mathbb{D},a,b}$.

This expression of the domain Markov property is more intuitive on the lattice in the discrete setting—because the interfaces are simple curves and because conditional probabilities have a much simpler definition—and it holds in many examples. It is vaguely related to the notion of locality in physics. The reader can check it straightforwardly for the exploration process. Equation (2.4) makes the domain Markov property plain for loop-erased random walks as well, whereas a direct proof using the original definition is more cumbersome.

2.4.3 Schramm's Argument

Our aim is to explore the interplay between conformal invariance and the domain Markov property of the family $\{\mathbf{P}_{\mathbb{D},a,b}\}$.

First, by conformal invariance, we may concentrate on the triple $(\mathbb{H}, 0, \infty)$. We choose a parameterisation of curves in $\Omega_{\mathbb{H},0,\infty}$ in such a way that the hull $\mathbb{K}_t := \mathbb{K}_{\gamma_t}$ associated with the initial segment $\gamma_{|0,t|} := \gamma_{|0,\gamma_t|}$ of $\gamma \in \Omega_{\mathbb{H},0,\infty}$ has capacity $2t$. Because of the underlying continuous curve γ, the growth of $\mathbb{K}_t$ is local, and the associated g_t satisfies a Loewner equation $\frac{\partial g_t(z)}{\partial t} = \frac{2}{g_t(z) - \xi_t}$ for some continuous function ξ_t. The probability $\mathbf{P}_{\mathbb{H},0,\infty}$ on $\Omega_{\mathbb{H},0,\infty}$ induces a random process on the set of initial segments $\gamma_{|0,t|}$, hence on the set of hulls $\mathbb{K}_t$, and on the set of continuous functions ξ_t.

Our next aim is to derive consequences for the stochastic process ξ_t of the domain Markov property and conformal invariance.

First for fixed $(\mathbb{H}, 0, \infty)$ there is a remnant of conformal invariance: dilatations. Hence for $\lambda > 0$, the hull $\frac{1}{\lambda} \mathbb{K}_{\lambda^2 t}$ must have the same distribution as a $\mathbb{K}_t$. The corresponding Loewner map is $\frac{1}{\lambda} g_{\lambda^2 t}(\lambda z)$, whose driving function is $\frac{1}{\lambda} \xi_{\lambda^2 t}$. Hence the processes ξ_t and $\frac{1}{\lambda} \xi_{\lambda^2 t}$ have the same law. We say that ξ_t has dimension $1/2$.

Given K_t, the domain Markov property states that $\gamma_{|t,\infty]}$ is distributed according to $\mathbf{P}_{\mathbb{D}\backslash K_t,\gamma_t,\infty}$. The conformal transformation $g_t(z) - \xi_t$ maps $\mathbb{D} \backslash K_t$ to $\mathbb{H}$, γ_t to 0 and ∞ to ∞. By conformal invariance, $g_t(\gamma_{|t,\infty]}) - \xi_t$ is distributed according to $\mathbf{P}_{\mathbb{H},0,\infty}$. In particular for $s \geq 0$ $g_t(\gamma_{|t,t+s]}) - \xi_t$ has the same distribution as a $\gamma_{|0,s]}$ hence is independent of $\gamma_{|0,t]}$. But the Loewner map for $g_t(\gamma_{|t,t+s]}) - \xi_t$ is $g_{s+t} \circ f_t(z + \xi_t) - \xi_t$ (remember f_t is the inverse of g_t), whose driving function is $\xi_{t+s} - \xi_t$. We infer that the random function $\xi.$ is such that for any $t, s \geq 0$, $\xi_{t+s} - \xi_t$ is independent of $\{\xi_{t'}\}$, $t' \in [0, t]$ and distributed like a ξ_s.

To resume our knowledge, the random process $\xi.$ has continuous samples, independent identically distributed increments and dimension $1/2$. By a deep general result, a random process with continuous samples and independent identically distributed increments is of the form $\sqrt{\kappa} B_t + \rho t$ for some non-negative κ and some real ρ. Obviously it has dimension $1/2$ if and only if $\rho = 0$.

To conclude, Schramm's argument shows that if a family of probabilities $\{\mathbf{P}_{\mathbb{D},a,b}\}$ on curves without crossing indexed by triples $(\mathbb{D}, a, b)$ is conformally invariant and has the domain Markov property, the law induced by $\mathbf{P}_{\mathbb{H},0,\infty}$ on initial hulls of capacity $2t$ by is described by a **stochastic Loewner evolution**

$$\frac{\partial g_t(z)}{\partial t} = \frac{2}{g_t(z) - \sqrt{\kappa} B_t} \tag{2.14}$$

for some $\kappa \geq 0$ and some normalised Brownian motion B_t. This process is often denoted by SLE_κ.

A priori, this does not show that each κ is realised via some family $\{\mathbf{P}_{\mathbb{D},a,b}\}$ (because the Loewner evolution deals wit hulls, not with curves).

2.4.4 Basic Properties

The first important property is a kind of converse to Schramm's result. If $\kappa \geq 0$ is a real number, and B_t a continuous realisation of a normalised Brownian motion, a deep theorem states that *the trace associated to the stochastic Loewner evolution equation* (2.14) *is almost surely a continuous curve joining* 0 *to* ∞. *This curve is simple and stays in* $\mathbb{H}$ *if* $\kappa \in [0, 4]$, *has double points and hits the real axis if* $\kappa \in]4, 8[$ *and is space filling if* $\kappa \in [8, +\infty[$.

At the time Schramm introduced stochastic Loewner evolutions, this very hard theorem was not known (he contributed to prove it later).

As explained before, a continuous trace cannot have crossings. Thus for any $\kappa \geq 0$, the stochastic Loewner evolution defines a probability measure $\mathbf{P}_\kappa$ on continuous curves without crossings joining 0 to ∞ in $\mathbb{H}$. This measure is scale-invariant. Hence, for each κ, conformal transformations can be used to define in a consistent way a family of probabilities $\{\mathbf{P}_{\mathbb{D},a,b}^\kappa\}$. This family is trivially conformally invariant, and it is easy to check that is satisfies the domain Markov property.

This finishes the complete classification.

Taking the existence of a curve for granted, the change of behaviour from simple curves to curves with double points at $\kappa = 4$ can be understood as follows. First, the necessary condition (negation of Eq. (2.12)) for the existence of multiple points is fulfilled for all values of κ, though in some kind of marginal way, for if $\xi_t = \sqrt{\kappa} B_t$ where B_t is a normalised Brownian motion, the law of the iterated logarithm states that, with probability one

$$\lim_{s \to t^-} \sup_{t' \in [s,t[} \frac{|\xi_t - \xi_{t'}|}{|t - t'|^{1/2} \ln \ln |t - t'|^{-1}} = \sqrt{2\kappa}.$$

So the stochastic Loewner source is wilder by a $\ln \ln |t - t'|^{-1}$ than the criterion. The fact that for $\kappa \leq 4$ the Loewner trace is a simple curve shows that, as should be expected, the criterion is only necessary, but not sufficient. Intuitively, Brownian motion is more singular than necessary, but for $\kappa \leq 4$ with too little correlation time to behave consistently for long enough periods to produce multiple points.

This fact is related to another well studied question: recurrence of Brownian motion. If space dimension $D = 1$, Brownian motion passes infinitely many times at any point. If $D = 2$, it passes infinitely many times in the any neighbourhood of any point, but not exactly at any given point, and if $D \geq 3$, it has a non-zero probability to remain at a given finite distance of any point. So dimension 2 is somehow a marginal case. Now let R_t be the norm of a d-dimensional Brownian motion. Assume $R_0 > 0$. One can show using stochastic calculus that $W_t := -R_t + \frac{D-1}{2} \int_0^t \frac{ds}{R_s}$ is a standard 1-dimensional Brownian motion. In this equation, D appears as an explicit parameter, and one can reverse the logic: given a standard 1-dimensional Brownian motion W_t what are the properties of R_t, called the d-dimensional **Bessel process** in mathematics. Setting $\kappa = 4/(d - 1)$ one sees that $X_t := \sqrt{\kappa}(R_t + W_t)$ satisfies the equation $\frac{dX_t}{dt} = \frac{2}{X_t - \sqrt{\kappa} W_t}$. This has two important consequences: first, one can indeed retrieve R_t from W_t by solving a differential equation and second, the Bessel process is essentially a stochastic Loewner evolutions but looking only at the boundary of $\mathbb{H}$. For general D, the Bessel processes behave with respect to visits to 0 just like the recurrence properties of Brownian motion for integer d suggest: the D-dimensional Bessel process hits the origin infinitely many times if $D < 2$, but never if $D \geq 2$. Equivalently, if $\kappa \leq 4$, $X_t - \sqrt{\kappa} W_t$ never vanishes, but vanishes infinitely many times if $\kappa > 4$. But we already know that the vanishing of $X_t - \sqrt{\kappa} W_t$ is the sign that the growing curve hits itself or the real axis.

Another very hard result is the fractal dimension: the measures $\mathbf{P}^\kappa_{\mathbb{D},a,b}$ is concentrated on curves with fractal dimension $\min\{1 + \kappa/8, 2\}$.

Recently, two important conjectures on SLE have been proven.

One of them is **reversibility**. The treatment of random curves by a Loewner evolution is quite asymmetric by definition. However interfaces between two points in physics (i.e. in statistical mechanics models) quite generally make no difference between the two ends. So it was conjectured very early that interfaces generated by an SLE process were reversible. One difficulty is with the parameterisation. Take an SLE sample in $\mathbb{H}$ from 0 to infinity, parameterise it with capacity. Apply the transformation $z \mapsto -1/z$ and parameterise the inverse sample with capacity. Now any point on the curve has two parameters attached to it. One of the troubles is that the relationship between the two parameters is extremely wild. Anyway, reversibility is now a theorem.

The second one is **duality**. Take an SLE_κ sample with $\kappa > 4$ and look at the boundary of $\mathbb{K}_t$. This is simple curve, and one can expect that its distribution is conformally invariant in some sense. So it is natural to ask if and how it fits in the SLE framework. It was conjectured by physicists that it is related in some sense to an $SLE_{16/\kappa}$, and that in particular it has dimension $1 + 2/\kappa$. Though this is correct, the precise recent theorem that gives an explicit description involves non-trivial extensions of SLE where the driving function is $\sqrt{16/\kappa}\,B_t$ plus some rather complicated drift terms.

2.4.5 Locality

Let us go back to percolation. Consider a domain $\mathbb{D}$ and three boundary points a, b, c such that $(\mathbb{D}, a, b)$ and $(\mathbb{D}, a, c)$ are hexagonal domains with admissible boundary conditions. On the boundary interval (b, c) the colours disagree. But both domains share the same inner hexagons, and the percolation samples are the same. For each configuration of inner coloured hexagons, it is clear that the exploration paths for $(\mathbb{D}, a, b)$ and $(\mathbb{D}, a, c)$ coïncide until they hit the boundary interval (b, c) for the first time. Hence the measures on interfaces stopped when they first hit (b, c) is the same for $(\mathbb{D}, a, b)$ and $(\mathbb{D}, a, c)$.

One could even go further and consider two hexagonal domains with admissible boundary conditions $(\mathbb{D}, a, b)$ and $(\mathbb{D}', a, b')$. Say that an hexagon in $\mathbb{D} \cap \mathbb{D}'$ is special if it is inner for $\mathbb{D}$ but not for $\mathbb{D}'$ or vice versa, or if it is a boundary hexagon for both, but with a different colour. By an analogous argument, the distribution for the exploration process started at a and stopped when it hits a special hexagon is the same for $(\mathbb{D}, a, b)$ and $(\mathbb{D}', a, b')$. This is called the **locality property**.

The particular case when $b = b'$ and $\mathbb{D}' \subset \mathbb{D}$ was mentioned before.

The notion of locality can also be formulated in the continuum. Let $\mathbb{L}$ be a hull in $\mathbb{D}$ bounded away from a and b. To each curve in $\Omega_{\mathbb{D},a,b}$ we can associate its smallest initial segment that hits the boundary of $\mathbb{L}$ (we take this initial segment to be the curve itself if it never hits $\mathbb{L}$). These initial segments form a system Σ of distinct representatives both in $\Omega_{\mathbb{D},a,b}$ and in $\Omega_{\mathbb{D}\setminus\mathbb{L},a,b}$. Thus both $\mathbf{P}_{\mathbb{D},a,b}$ and $\mathbf{P}_{\mathbb{D}\setminus\mathbb{L},a,b}$ induce a probability measure on Σ. The property of **locality** is the statement that these two measures coincide. In a more mundane way, if $\mathbb{L}$ is a hull in $\mathbb{D}$ bounded away from a and b, the distribution of curves up to the first hitting of $\mathbb{L}$ are the same in $\mathbb{D}$ and in $\mathbb{D} \setminus \mathbb{L}$.

Stochastic calculus can be used to show that the family $\{\mathbf{P}^{\kappa=6}_{\mathbb{D},a,b}\}$ is the only one to have the locality property. Let us note that it is no surprise that a value of κ satisfying locality is > 4. Indeed, if $\kappa \leq 4$, the traces are simple curves that do not hit the boundary. Then no trace touches $\mathbb{L}$ for $\mathbf{P}_{\mathbb{D}\setminus\mathbb{L},a,b}$, but hitting $\mathbb{L}$ for $\mathbf{P}_{\mathbb{D},a,b}$ has a finite probability if $\mathbb{L}$ is non-trivial, so that the supports of the two probability measures induced on Σ are not the same.

Though the amount of mathematical machinery is significantly higher than in the rest of these notes, let us give the outline of a computation that shows that *only SLE_6 can have the locality property*.

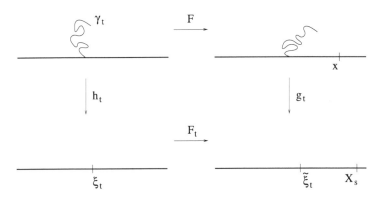

Fig. 2.18 The basic commutative diagram. (*Left*) Half plane, from 0 to infinity. (*Right*) Half plane, from 0 to x

One compares stochastic Loewner evolution in $\mathbb{H}$ from 0 to ∞ to stochastic Loewner evolution in $\mathbb{H}$ from 0 to x where x is any boundary point. By conformal invariance, any map F from $\mathbb{H}$ to $\mathbb{H}$ sending 0 to 0 and ∞ to x is such that the image measure $\mathbf{P}_{\mathbb{H},0,\infty} \circ F^{-1}$ is $\mathbf{P}_{\mathbb{H},0,x}$. We fix such an F. Uniformise SLE in the upper half-plane from 0 to ∞ with trace $\gamma_{[0,t]}$ up to time t by $h_t : \mathbb{H}\backslash\gamma_{[0,t]} \to \mathbb{H}$ in the hydrodynamical normalisation:

$$\frac{dh_t(w)}{dt} = \frac{2}{h_t(w) - \xi_t}, \quad \xi_t = \sqrt{\kappa}B_t.$$

Let g_t be the uniformising map $g_t : \mathbb{H}\backslash F(\gamma_{[0,t]}) \to \mathbb{H}$ in the hydrodynamical normalisation as above except that the time parameterisation is not at our disposal:

$$\frac{dg_t(w)}{dt} = \frac{2a_t}{g_t(w) - \tilde{\xi}_t}.$$

Finally let $F_t : \mathbb{H} \to \mathbb{H}$ be the map "closing the square",

$$g_t \circ F = F_t \circ h_t.$$

The commutative diagram Fig. 2.18 summarises the situation.

Take the time derivative of $g_t \circ F = F_t \circ h_t$ and afterwards substitute z for h_t to get

$$\frac{2a_t}{F_t(z) - \tilde{\xi}_t} = \frac{dF_t(z)}{dt} + F_t'(z)\frac{2}{z - \xi_t}.$$

Now $F_t(z)$ is non-singular at $z = \xi_t$ so that the pole on the right-hand side is cancelled by a pole on the left-hand side:

$$\frac{2a_t}{F_t(z) - \tilde{\xi}_t} = F_t'(\xi_t)\frac{2}{w - \xi_t} + O(1) \quad \text{when } w \to \xi_t,$$

leading to $\tilde{\xi}_t = F_t(\xi_t)$ and $a_t = F'_t(\xi_t)^2$. Continuing the expansion one step further yields

$$-a_t F''_t(\xi_t)/F'_t(\xi_t)^2 = \frac{\mathrm{d}F_t}{\mathrm{d}t}(\xi_t) + 2F''_t(\xi_t) \quad \text{i.e.} \quad \frac{\mathrm{d}F_t}{\mathrm{d}t}(\xi_t) = -3F''_t(\xi_t).$$

Now use Itô's formula to get $\mathrm{d}\tilde{\xi}_t = \mathrm{d}(F_t(\xi_t)) = (\mathrm{d}F_t)(\xi_t) + F'_t(\xi_t)\,\mathrm{d}\xi_t + \frac{\kappa}{2}F''_t(\xi_t)\,\mathrm{d}t$, i.e.

$$\mathrm{d}\tilde{\xi}_t = F'_t(\xi_t)\,\mathrm{d}\xi_t + (\kappa/2 - 3)F''_t(\xi_t)\,\mathrm{d}t.$$

Let us analyse this equation naïvely. As usual in Itô's theory, consider $\int_0^t F'_u(\xi_u)\,\mathrm{d}\xi_u \simeq \sum_{n=0}^{N-1} F'_{tn/N}(\xi_{tn/N})(\xi_{t(n+1)/N} - \xi_{tn/N})$. If ξ is known up to time tn/N, the random variable $F'_{tn/N}(\xi_{tn/N})(\xi_{t(n+1)/N} - \xi_{tn/N})$ is Gaussian with mean 0 and variance $F'_{tn/N}(\xi_{tn/N})^2(\kappa t/N)$. Note that the increments $(\xi_{t(n+1)/N} - \xi_{tn/N})$ are all independent, and the scale $F'_{tn/N}(\xi_{tn/N})$ depends solely on the past. This suggests that if we would count time with a different scale, $\int_0^t F'_u(\xi_u)\,\mathrm{d}\xi_u$ would become a sum of independent Gaussian random variables, and by inspection the right time scale is $s(t) := \int_0^t F'_u(\xi_u)^2\,\mathrm{d}u$. This hand-waving argument is in fact confirmed by a theorem of Itô: if $\sqrt{\kappa}W_{s(t)} := \int_0^t F'_u(\xi_u)\,\mathrm{d}\xi_u$ then W_s is a standard Brownian motion with parameter s. Setting $\hat{\xi}_{s(t)} := \tilde{\xi}_t$ we get that $\hat{\xi}_{s(t)} := F_t(\xi_t)$ so

$$\mathrm{d}\hat{\xi}_s = \sqrt{\kappa}\,\mathrm{d}W_s + (\kappa/2 - 3)\frac{F''_{t(s)}(\xi_{t(s)})}{F'_{t(s)}(\xi_{t(s)})^2}\,\mathrm{d}s.$$

Observe that the time change is also exactly the one needed to turn $a_t = F'_t(\xi_t)^2$ into the constant 1. So if we set $\hat{h}_{s(t)} := g_t$ we find

$$\frac{\mathrm{d}\hat{h}_s(w)}{\mathrm{d}s} = \frac{2}{\hat{h}_s(w) - \hat{\xi}_s}.$$

This is enough to conclude that SLE$_\kappa$ from 0 to ∞ is described by that same equation as SLE$_\kappa$ from 0 to x if and only if $\kappa = 6$. As in the discrete case, this leads to the fact that at $\kappa = 6$ (and for no other value of κ) the measures $\mathbf{P}_{\mathbb{H},0,x}$ (resp. $\mathbf{P}_{\mathbb{H},0,y}$) on traces from 0 to x (resp. to y) induce the same measure on traces starting and 0 and stopped when they hit the interval $]x, y[$.

In the simple case at hand, one can prove straightforwardly that

$$\frac{F''_{t(s)}(\xi_{t(s)})}{F'_{t(s)}(\xi_{t(s)})^2} = \frac{2}{\tilde{\xi}_t - g_t(x)}.$$

Defining $\hat{X}_s := \hat{h}_s(x)$ we get finally:

$$\mathrm{d}\hat{\xi}_s = \sqrt{\kappa}\,\mathrm{d}W_s + (\kappa - 6)\frac{\mathrm{d}s}{\hat{\xi}_s - \hat{X}_s}, \qquad \frac{\mathrm{d}\hat{X}_s}{\mathrm{d}s} = \frac{2}{\hat{X}_s - \hat{\xi}_s},$$

so what looked initially like a non-local stochastic differential equation is in fact a local system of coupled stochastic differential equations.

2.5 Relation with Conformal Field-Theory

In this section, we shall crudely outline the relation between SLE and CFT. The discussion will be rather informal.

This is the only part of these notes were the author can claim to have made a contribution.

2.5.1 Motivation

The starting point is the following. If g_t is the Loewner map for SLE from 0 to ∞, we define $h_t := g_t - \xi_t$. Then h_t satisfies the stochastic differential equation

$$\mathrm{d}h_t(z) = \frac{2}{h_t(z)}\mathrm{d}t - \mathrm{d}\xi_t.$$

We can rephrase this trivially by setting $Z_t := h_t(z)$, so that

$$\mathrm{d}Z_t = \frac{2}{Z_t}\,\mathrm{d}t - \mathrm{d}\xi_t, \qquad Z_0 = z.$$

This describes the motion of particles in a time-dependent vector field v such that

$$v\,\mathrm{d}t = \mathrm{d}t\frac{2}{z}\partial_z - \mathrm{d}\xi_t\partial_z.$$

For each $n \in \mathbb{Z}$, the generator of the transformation $z \to z + \varepsilon z^{n+1}$ is $\ell_n := -z^{n+1}\partial_z$ so that

$$v\,\mathrm{d}t = -2\ell_{-2}\,\mathrm{d}t + \ell_{-1}\,\mathrm{d}\xi_t.$$

Now, if we consider a function of Z_t, say f, we can apply Itô's formula (or any of the techniques that are more familiar to physicists) to get the variation of the expectation value of $f(Z_t)$ namely

$$\frac{\mathrm{d}}{\mathrm{d}t}E\big(f(Z_t)\big)\bigg|_{t=0} = \left(\frac{\kappa}{2}\partial_z^2 + \frac{2}{z}\partial_z\right)f(z).$$

The operator acting on the right-hand side is $\frac{\kappa}{2}\ell_{-1}^2 - 2\ell_{-2}$. Functions in the kernel of this operator describe observables $f(Z_t)$ with a time independent expectation, which in this simple case are also examples of more general probabilistic objects known as martingales.

But this differential operator is also well-known in conformal field-theory.

2.5.2 A Crash Course in Boundary CFT (BCFT)

The following is a totally unfair presentation of CFT. The first chapter in this volume, and references therein, should be consulted for any serious study.

In CFT the operators $\ell_n = -z^{n+1}\partial_z$, which satisfy the following commutation relations:

$$[\ell_m, \ell_n] = (m - n)\ell_{m+n}, \tag{2.15}$$

are represented by operators L_n which satisfy the **Virasoro algebra** $\mathfrak{vir}$:

$$[L_m, L_n] = (m - n)L_{m+n} + \frac{c}{12}m(m^2 - 1)\delta_{m+n;0} \tag{2.16}$$

with the Virasoro **central charge** c, which commutes with all other generators (and is often viewed simply as a scalar multiple of the identity, though this is strictly true only in irreducible representations). The Virasoro algebra is a (in fact the only non-trivial) central extension of the algebra (2.15). This means that for non-zero c the quantum field-theory implements only a projective representation of the Lie algebra (2.15) of infinitesimal conformal transformations.

The states of a CFT are organised[5] in **highest-weight representations** of $\mathfrak{vir}$. They are characterised by the facts that first, they possess a **highest-weight vector** i.e. a vector $|\Delta\rangle$ such that $L_n|\Delta\rangle = 0$ for $n > 0$ and $L_0|\Delta\rangle = \Delta|\Delta\rangle$ and second, all states in the representation of $\mathfrak{vir}$ are linear combinations of states obtained by acting on $|\Delta\rangle$ repeatedly with Virasoro generators. The parameter Δ is called the **conformal weight** of the representation.

Due to these conditions and the commutation relations, the most general state in the representation is a finite linear combination of eigenstates of L_0 of eigenvalue $m + \Delta$ for $m = 0, 1, \ldots$. An application of the **Poincaré-Birkhoff-Witt theorem** yields that these eigenstates with eigenvalue $m + \Delta$ can be written in the form $L_{-n_1}L_{-n_2}\ldots L_{-n_k}|\Delta\rangle$ where $n_1 \geq n_2 \geq \ldots \geq n_k \geq 1$ and $n_1 + n_2 + \ldots + n_k = m$. In generic cases, these states are linearly independent. This is always true in a certain type of representations called **Verma modules** (and this can be taken as a poor man's definition/characterisation). But for certain special values of the pair (c, Δ), Verma modules contain submodules and the corresponding quotients still satisfy the highest-weight condition. A bit more on this later.

In what follows, we shall mostly deal with boundary observables. To each state is associated a boundary operator, and the boundary operator $\varphi_\Delta(x)$ associated to $|\Delta\rangle$ for a CFT in the upper half-plane $\mathbb{H}$ (so that $x \in \mathbb{R}$) is called a **boundary primary operator** and satisfies

$$\left[L_n, \varphi_\Delta(x)\right] = \left(x^{n+1}\frac{\mathrm{d}}{\mathrm{d}x} + \Delta x^n\right)\varphi_h(x), \tag{2.17}$$

which is the operator infinitesimal version of the rule $\varphi_\Delta(x) \mapsto \varphi_\Delta(f(x))|f'(x)|^\Delta$ under a conformal transformation f from a domain to another one, with the proviso that f should have a tangential derivative at the boundary point x. The state $|\Omega\rangle$ is associated to the identity operator, and $|\Delta\rangle = \varphi_\Delta(x)|\Omega\rangle$.

---

[5]This property of CFTs is in fact not general enough to cover all interesting cases, in particular for the SLE-CFT correspondence. However, the author is not aware of a simple and fully general definition of a CFT and decided to remain in the simplest setting.

2.5.3 Martingales and Singular Vectors

If one elaborates on the crude discussion of the $O(n)$ model in Sect. 2.2.2.3, one is lead to conjecture that the interface is created by inserting some boundary changing operators, say ψ sitting at the beginning and at the end of the interface. As noted there, this does not really imply that one obtains in this way a "product of local observables". If $\langle\,\rangle$ denotes the average in the system without interface, we expect that the average of the observable $\mathcal{O}$ in the presence of an interface joining points a and b on the boundary will be

$$\langle\mathcal{O}\rangle_{a,b} := \frac{\langle\mathcal{O}\psi(a)\psi(b)\rangle}{\langle\psi(a)\psi(b)\rangle}. \tag{2.18}$$

This formula has the correct covariance properties under conformal transformations if and only if ψ transforms in a homogeneous way, i.e. is a density associated to some highest weight in a representation of the Virasoro algebra. We would like to understand which one.

The crucial observation is the following: if one computes the average of $\mathcal{O}$ for a fixed position of the interface and then averages over the position of the interface with the correct measure, one retrieves $\langle\mathcal{O}\rangle_{a,b}$. In the case when the interface is described by a growth process like SLE, this can be rephrased by saying that *observables are functionals of the Loewner flow whose average is time-independent.* Together with the domain Markov property, this means that *these functionals are* **martingales**. The discussion in the last section gives a detailed proof of this statement on the lattice.

The simplest functional has been considered above and leads to consider the kernel of $\frac{\kappa}{2}\ell_{-1}^2 - 2\ell_{-2}$.

So the natural question is: is it possible to have

$$\left(\frac{\kappa}{2}L_{-1}^2 - 2L_{-2}\right)|h\rangle = 0?$$

The answer is that it may occur if and only if

$$c = \frac{(6-\kappa)(3\kappa-8)}{2\kappa} = 1 - 6\frac{(\kappa-4)^2}{4\kappa}$$

and $\Delta = \frac{6-\kappa}{2\kappa}$.

One can make this heuristic discussion more rigorous and get:

- SLEs with parameter κ describe interfaces in CFTs of Virasoro central charge

$$c_\kappa = \frac{(6-\kappa)(3\kappa-8)}{2\kappa} = 1 - 6\frac{(\kappa-4)^2}{4\kappa}. \tag{2.19}$$

Notice that c_κ is always less than 1 and is invariant under the duality $\kappa \leftrightarrow 16/\kappa$.

- The boundary conformal operator $\psi(x)$ implementing the change of boundary condition at the point on which the interface emerges has scaling dimension[6]

$$\Delta_{1;2} = \frac{6-\kappa}{2\kappa}. \tag{2.20}$$

It is a Virasoro primary operator degenerate at level two. In the CFT literature, this operator is often denoted by $\psi_{1;2}$.

The notation makes references to the so-called **Kac's labels** $(r;s)$ which parameterise an important family of representations, with conformal weight $\Delta_{r;s} = \frac{(\kappa r - 4s)^2 - (\kappa - 4)^2}{16\kappa}$. When r and s are positive integers, the corresponding Verma module contains non-trivial submodules. For generic κ, there is only one submodule, which is responsible for the fact that a CFT correlation functions of an operator $\psi_{r,s}$ corresponding to the quotiented Virasoro representation satisfies a differential equation of order rs. The case $(r;s) = (1;1)$ is the one of the identity operator. The case $(r;s) = (1;2)$ is at the heart of the **SLE-CFT correspondence**. When κ is a rational number, some conformal weights can be written as $\Delta_{r;s}$ for different values of $(r;s)$, leading to a more complicated submodule structure and to Virasoro minimal models. Note that some non-degenerate pairs (r,s) play a role in the CFTs of SLEs. For instance, a bulk operator with conformal weights $(\Delta_{0;n/2}, \Delta_{0;n/2})$ represents the insertion of n interfaces from a point in the bulk. In fact this is why, for instance, $2 - 2\Delta_{0;1} = (\kappa + 8)/8$ is the fractal dimension of SLE (for $\kappa \le 8$). Note that $(0; n/2)$ is never in the Kac's table, and $n/2$ is not even an integer for odd n.

2.5.4 Two Examples

2.5.4.1 Hitting Probability

If $0 < u < v$, one can ask for the probability that an SLE trace from 0 to ∞ in $\mathbb{H}$ hits $[u, v]$, see Fig. 2.19. Denote this by $p(u, v)$. Suppose we let the trace grow for a certain time t and ask, then, for the same probability. If the trace is described by g_t (in the hydrodynamical normalisation), by conformal invariance, this is $p(g_t(u) - \xi_t, g_t(v) - \xi_t)$. If this is averaged over the position of the trace up to time t, $p(u, v)$ is retrieved. Writing

$$\frac{d}{dt} E\big(p\big(g_t(u) - \xi_t, g_t(v) - \xi_t\big)\big)\bigg|_{t=0} = 0$$

yields a partial differential equation, obtained again routinely either via Itô's formula or any method more familiar to physicists:

$$\left(\frac{2}{u}\partial_u + \frac{2}{v}\partial_v + \frac{\kappa}{2}(\partial_u + \partial_v)^2\right)p(u, v) = 0.$$

---

[6]For a boundary operator, the conformal weight is exactly the scaling dimension. Bulk operators have (possibly different) left and right conformal weights. Their sum is the scaling dimension and their difference is the spin.

Fig. 2.19 Hitting
configuration

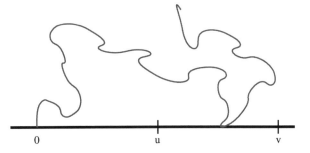

By scale invariance, $p(u, v)$ depends only on the ratio u/v and inserting boundary
conditions shows that $p(u, v) = 0$ for $\kappa \leq 4$, $p(u, v) = 1$ for $\kappa \geq 8$. One also gets
an explicit formula for $\kappa \in \,]4, 8[$. Setting $s = u/v$,

$$1 - p(u, v) = \frac{s^{\frac{\kappa-4}{\kappa}} \Gamma(\frac{4}{\kappa})}{\Gamma(\frac{\kappa-4}{\kappa}) \Gamma(\frac{8-\kappa}{\kappa})} \int_0^1 d\sigma \, \sigma^{-\frac{4}{\kappa}} (1 - s\sigma)^{2\frac{4-\kappa}{\kappa}}.$$

On the CFT side, on finds that $1 - p(u, v) = \langle \psi_{1;2}(\infty)\varphi_0(v)\varphi_0(u)\psi_{1;2}(0)\rangle$ were
the operator φ_0 is a primary operator of weight 0 and appropriate conformal blocks
are chosen. Starting from

$$\left(\frac{\kappa}{2}L_{-1}^2 - 2L_{-2}\right)|\psi_{1;2}\rangle = 0, \qquad \langle \psi_{1;2}|L_n = 0 \quad \text{for } n = -1, -2, \ldots$$

plus the commutation relations between Virasoro generators and primary operators
(2.17), it is a routine CFT computation to show, that $\langle \psi_{1;2}(\infty)\varphi_0(v)\varphi_0(u)\psi_{1;2}(0)\rangle$
satisfies the same partial differential equation as the one obtained above by proba-
bilistic arguments. The boundary conditions are that it should take value 0 for $u \to 0$
and value 1 at $u \to v$. At least, this is what the probabilistic computation shows. To
motivate this from the CFT viewpoint, we argue as follows.

If the trace hits between u and v, $g_t(u)$ will be close to ξ_t for t close to the hitting
time, so that an OPE of $\varphi_0(g_t(u))$ with $\psi_{1;2}(\xi_t)$ will be relevant. Due to the existence
of the null vector $(\frac{\kappa}{2}L_{-1}^2 - 2L_{-2})|\Delta_{1;2}\rangle$, only two conformal families are possible
in the OPE, one with dimension $\Delta_{1;2} = \frac{6-\kappa}{2\kappa}$ and one with dimension $\Delta_{1;0} = \frac{\kappa-2}{\kappa}$.
So

$$\varphi_0(x)\psi_{1;2}(0) \simeq C_{1;2}\big(\psi_{1;2}(0) + O(x)\big) + C_{1;0}x^{\frac{\kappa-4}{\kappa}}\big(\psi_{1;0}(0) + O(x)\big).$$

As hitting is forbidden in $1 - p(u, v)$, we have to choose a channel which gives a
vanishing small contribution at small x, i.e. κ must be > 4 and we have to pick only
the $\psi_{1;0}$ channel, i.e. $C_{1;2} = 0$. This fixes the correlation function up to multiplication by a constant.

On the other hand, if the trace does not hit between u and v, the first time it hits
after v, $g_t(u)$ and $g_t(v)$ will be close to ξ_t. But, as observed in the explicit examples
of hittings, when a trace hits the boundary, $g_t(u)$ and $g_t(v)$ come close to each other
at an even faster rate, so that an OPE between $\varphi_0(g_t(u))$ and $\varphi_0(g_t(v))$ is relevant.
This time, there is no a priori restrictions on the possible operators, but only two

conformal families have a non-vanishing three point function with the two $\psi_{1;2}$'s: one with dimension $\Delta_{1;1} = 0$ and one with dimension $\Delta_{1;3} = \frac{8-\kappa}{\kappa}$. So one picks only

$$\varphi_0(x)\varphi_0(y) \simeq C_{1;1}\big(1 + O(x-y)\big) + C_{1;3}(x-y)^{\frac{8-\kappa}{\kappa}}\big(\psi_{1;3}(y) + O(x-y)\big) + \dots.$$

Now one can check that if $C_{1;2} = 0$ and $C_{1;0} \neq 0$ then $C_{1;3} \neq 0$ as well, so the correlation function is bounded only if $\kappa < 8$ (at $\kappa = 8$, their would be logarithms in fact).

So the block structure is dictated by the operator product expansion. In this way, the OPE encodes nicely the different phases of SLE: only for $\kappa \in]4, 8[$ is there a non-trivial hitting probability, and hitting the boundary is represented by insertion of the operator $\psi_{1;3}$.

2.5.4.2 Partition Functions

As argued before, the partition function for chordal SLE in $\mathbb{D}$ from a to b (a and b are two boundary points of $\mathbb{D}$) is quite simple:

$$Z_{\mathbb{D}}(a,b) = \big\langle \psi_{h_{1,2}}(a)\psi_{h_{1,2}}(b)\big\rangle_{\mathbb{D}}.$$

To be sure, we should multiply the right-hand side by the partition function without interface. This factor depends on $\mathbb{D}$ but plays no role in the following arguments and we omit it. Conformal invariance relates correlation functions in different domains: if $g : \mathbb{D} \to \tilde{\mathbb{D}}$ is a conformal representation, one finds, using the behaviour of boundary primary operators under conformal transformations:

$$\big\langle \psi_{h_{1,2}}(a)\psi_{h_{1,2}}(b)\big\rangle_{\mathbb{D}} = \big\langle \psi_{h_{1,2}}(g(a))\psi_{h_{1,2}}(g(b))\big\rangle_{\tilde{\mathbb{D}}}\big|g'(a)\big|^{\Delta_{1;2}}\big|g'(b)\big|^{\Delta_{1;2}},$$

where $\Delta_{1;2} = \frac{6-\kappa}{2\kappa}$.

In particular, taking $\mathbb{D} = \tilde{\mathbb{D}} = \mathbb{H}$ (the upper-half plane, a and b are real numbers in that case) and $g(z) = \frac{z-a}{pz+(1-p)b-a}$ where p is an arbitrary real parameter, $p \neq 1$, one gets that

$$\big\langle \psi_{\Delta_{1,2}}(a)\psi_{\Delta_{1,2}}(b)\big\rangle_{\mathbb{H}} = \big\langle \psi_{\Delta_{1,2}}(0)\psi_{\Delta_{1,2}}(1)\big\rangle_{\mathbb{H}}|a-b|^{-2\Delta_{1;2}}.$$

Taking $\mathbb{D}$ to be arbitrary again, but keeping $\tilde{\mathbb{D}} = \mathbb{H}$ one gets

$$\big\langle \psi_{\Delta_{1,2}}(a)\psi_{\Delta_{1,2}}(b)\big\rangle_{\mathbb{D}} = \big\langle \psi_{\Delta_{1,2}}(g(a))\psi_{\Delta_{1,2}}(g(b))\big\rangle_{\mathbb{H}}\big|g'(a)\big|^{\Delta_{1;2}}\big|g'(b)\big|^{\Delta_{1;2}}$$

$$= \big\langle \psi_{\Delta_{1,2}}(0)\psi_{\Delta_{1,2}}(1)\big\rangle_{\mathbb{H}}\left|\frac{g'(a)g'(b)}{(g(a)-g(b))^2}\right|^{\frac{(6-\kappa)}{2\kappa}}.$$

As expected, these formulæ are singular if $\mathbb{D}$ is not smooth at a or b. However, comparing the cases $b = x$ and $b = y$, one is led to

$$\frac{Z_{\mathbb{D}}(a,y)}{Z_{\mathbb{D}}(a,x)} = \frac{\big\langle \psi_{\Delta_{1,2}}(a)\psi_{\Delta_{1,2}}(y)\big\rangle_{\mathbb{D}}}{\big\langle \psi_{\Delta_{1,2}}(a)\psi_{\Delta_{1,2}}(x)\big\rangle_{\mathbb{D}}}$$

$$= \left|\frac{g'(y)(g(a)-g(x))^2}{g'(x)(g(a)-g(y))^2}\right|^{\frac{(6-\kappa)}{2\kappa}}.$$

This formula has two remarkable features:

- the normalising constants in the partition functions or in the two point functions have cancelled between the numerator and denominator,
- the derivative $g'(a)$ has cancelled between the numerator and denominator.

Hence, this formula is normalised in an absolute way and makes sense even if $\mathbb{D}$ is not smooth at a. We want to apply it in such a situation.

Consider chordal SLE in $\mathbb{H}$ from 0 to x. Let $\mathbb{D} = \mathbb{H}_t$ be the domain obtained by removing from $\mathbb{H}$ the SLE hull with the SLE hull at time t, and let $g_t : \mathbb{H}_t \to \mathbb{H}$ be the uniformising map in the hydrodynamical normalisation, mapping the tip γ_t of the SLE hull to ξ_t. Modulo some changes in notation, we have seen at the end of Sect. 2.4.5 that

$$\mathrm{d}\xi_t = \sqrt{\kappa}\,\mathrm{d}B_t + (\kappa - 6)\frac{\mathrm{d}t}{\xi_t - g_t(x)}, \qquad \frac{\mathrm{d}g_t(x)}{\mathrm{d}t} = \frac{2}{g_t(x) - \xi_t}, \qquad (2.21)$$

where B_t is a standard Brownian motion.

Taking $a = \gamma_t$ (a point at which $\mathbb{D} = \mathbb{H}_t$ is certainly not smooth) we obtain for the ratio of partition functions:

$$\frac{Z_{\mathbb{H}_t}(\gamma_t, y)}{Z_{\mathbb{H}_t}(\gamma_t, x)} = \left| \frac{g'_t(y)(\xi_t - g_t(x))^2}{g'_t(x)(\xi_t - g_t(y))^2} \right|^{\frac{(6-\kappa)}{2\kappa}}.$$

We claim that this is a martingale for SLE in $\mathbb{H}$ from 0 to x, i.e. for the stochastic differential system of Eqs. (2.21). The proof is a good exercise with **Itô's formula** that we really recommend to the reader. Therefore we give some details. Note that all factors in the ratio are real, so that the modulus is only fixing a possible sign. Hence we can forget about the modulus in the computation. In the following we set $X_t := g_t(x) - \xi_t$ and $Y_t := g_t(y) - \xi_t$.

As $\frac{\mathrm{d}g_t(z)}{\mathrm{d}t} = \frac{2}{g_t(z)-\xi_t}$, taking the z derivative (denoted with a $'$) yields

$$\mathrm{d}g'_t(z) = \frac{-2g'_t(z)\,\mathrm{d}t}{(g_t(z) - \xi_t)^2},$$

which gives

$$\mathrm{d}\frac{g'_t(y)}{g'_t(x)} = -2(X_t - Y_t)\frac{g'_t(y)}{g'_t(x)}\frac{(X_t + Y_t)}{X_t^2 Y_t^2}\,\mathrm{d}t.$$

Also

$$\mathrm{d}\big(\xi_t - g_t(x)\big) = \mathrm{d}\xi_t - \mathrm{d}g_t(x) = \mathrm{d}\xi_t - \frac{2\,\mathrm{d}t}{X_t}.$$

Now Itô's formula yields

$$\mathrm{d}\frac{1}{\xi_t - g_t(y)} = -\frac{\mathrm{d}(\xi_t - g_t(y))}{(\xi_t - g_t(y))^2} + \frac{(\mathrm{d}(\xi_t - g_t(y)))^2}{(\xi_t - g_t(y))^3}$$

$$= -\frac{\mathrm{d}\xi_t}{Y_t^2} - (\kappa - 2)\frac{\mathrm{d}t}{Y_t^3}$$

where we used Itô's rule $dB_t^2 = dt$, $dB_t\, dt = dt\, dt = 0$ to obtain $(d(\xi_t - g_t(y)))^2 = \kappa\, dt$, a consequence of (2.21). Another use of Itô's formula yields

$$d\frac{\xi_t - g_t(x)}{\xi_t - g_t(y)} = \frac{1}{\xi_t - g_t(y)}d\big(\xi_t - g_t(x)\big) + \big(\xi_t - g_t(x)\big)\,d\frac{1}{\xi_t - g_t(y)}$$
$$+ \left(d\big(\xi_t - g_t(x)\big)\,d\frac{1}{\xi_t - g_t(y)}\right)$$
$$= (X_t - Y_t)\left(\frac{1}{Y_t^2}\,d\xi_t + \frac{-2Y_t + (\kappa - 2)X_t}{X_t Y_t^3}\,dt\right).$$

Then

$$d\left(\frac{\xi_t - g_t(x)}{\xi_t - g_t(y)}\right)^2 = (X_t - Y_t)\left(\frac{2X_t}{Y_t^3}\,d\xi_t + \frac{-4(X_t + Y_t) + \kappa(3X_t - Y_t)}{Y_t^4}\,dt\right).$$

Putting all this together and setting $S_t := \frac{g_t'(y)(\xi_t - X_t)^2}{g_t'(x)(\xi_t - Y_t)^2}$, we get

$$dS_t = \frac{g_t'(y)}{g_t'(x)}(X_t - Y_t)\left(\frac{2X_t}{Y_t^3}\,d\xi_t + \frac{-6(X_t + Y_t) + \kappa(3X_t - Y_t)}{Y_t^4}\,dt\right),$$

or better

$$\frac{dS_t}{S_t} = (X_t - Y_t)\left(\frac{2}{X_t Y_t}\,d\xi_t + \frac{-6(X_t + Y_t) + \kappa(3X_t - Y_t)}{X_t^2 Y_t^2}\,dt\right).$$

Itô's formula applied once again gives, for any exponent α

$$\frac{dS_t^\alpha}{S_t^\alpha} = \alpha\frac{dS_t}{S_t} + \frac{1}{2}\alpha(\alpha - 1)\left(\frac{dS_t}{S_t}\right)^2.$$

Hence

$$\frac{dS_t^\alpha}{S_t^\alpha} = \alpha(X_t - Y_t)$$
$$\times\left(\frac{2}{X_t Y_t}\,d\xi_t + \frac{-6(X_t + Y_t) + \kappa(3X_t - Y_t) + 2\kappa(\alpha - 1)(X_t - Y_t)}{X_t^2 Y_t^2}\,dt\right).$$

Recalling that $d\xi_t = \sqrt{\kappa}\,dB_t + (\kappa - 6)\frac{dt}{\xi_t - g_t(x)} = \sqrt{\kappa}\,dB_t + (6 - \kappa)\frac{dt}{X_t}$, we get at last:

$$\frac{dS_t^\alpha}{S_t^\alpha} = \alpha(X_t - Y_t)\left(\frac{2}{X_t Y_t}\sqrt{\kappa}\,dB_t + \frac{(\kappa(1 + 2\alpha) - 6)(X_t - Y_t)}{X_t^2 Y_t^2}\,dt\right). \qquad (2.22)$$

The drift term vanishes if and only if $\kappa(1 + 2\alpha) - 6 = 0$, i.e. $\alpha = h_{1;2} = (6 - \kappa)/(2\kappa)$. In particular, this proves that $\frac{Z_{\mathbb{H}_t}(\gamma_t, y)}{Z_{\mathbb{H}_t}(\gamma_t, x)} = |S_t|^{h_{1;2}}$ is a (local) martingale, i.e. that its Itô derivative contains no drift term.

The fact that the drift term in $d\xi_t$ is $(6 - \kappa)\frac{dt}{X_t}$ is crucial, and our computation shows that we could recover this drift term uniquely if we knew in advance that $\frac{Z_{\mathbb{H}_t}(\gamma_t, y)}{Z_{\mathbb{H}_t}(\gamma_t, x)}$ is a martingale. Moreover, if we were to take $\alpha \neq h_{1;2}$, S_t^α would be a martingale for no choice of drift term in $d\xi_t$. These two properties should convince us that the martingale property of $\frac{Z_{\mathbb{H}_t}(\gamma_t, y)}{Z_{\mathbb{H}_t}(\gamma_t, x)}$ has little to do with chance.

In the next subsection, we give the fundamental (but totally elementary) reason why *ratios of partition functions (and in particular correlation functions) must be martingales* and give a precise meaning to this vague statement. A good part of the argument relies on a very general tautological double counting argument which then is specialised to the SLE-CFT correspondence.

To close this discussion, let us note that the above computation can be abstracted as follows: The stochastic process

$$R_t = \frac{Z_{\mathbb{H}_t}(\gamma_t, y)}{Z_{\mathbb{H}_t}(\gamma_t, x)} \tag{2.23}$$

is a **martingale** if

$$d\big(g_t(\gamma_t)\big) = \sqrt{\kappa}\, dB_t + \kappa \partial_b \big(\ln Z_{\mathbb{H}}(a, b)\big)\big|_{a=g_t(\gamma_t),\, b=g_t(x)}\, dt. \tag{2.24}$$

This equation which states that the drift is the variation of the free energy with respect to a parameter is particularly nice, and in fact totally general. For instance, it is at the heart of the definition of multiples SLEs.

2.5.5 Conformal SLE Martingales

In this last subsection, we explain why *ratios of (conditional) partition functions of models of statistical mechanics (and in particular correlation functions) are martingales for appropriate stochastic processes*. This result is at the heart of the CFT approach to SLEs as outlined in the previous section. We first give the argument, which involves some technical hypotheses, and then motivate these hypotheses and their consequences in the example of interfaces. To keep the discussion simple and self-contained, we concentrate on finite or countable configuration spaces. Again, we use some terminology from probability theory (σ-algebra, martingale, filtration, ...), but try to keep it to a bare minimum. We start with some general abstract definitions.

Recall that a **partition** of a set $\mathscr{C}$ is a subset $\mathscr{Q}$ of $2^{\mathscr{C}} \setminus \{\emptyset\}$ such that each $x \in \mathscr{C}$ belongs to exactly one element of $\mathscr{Q}$. We say that a partition $\mathscr{Q}'$ is **finer** than a partition $\mathscr{Q}$, or equivalently that a partition $\mathscr{Q}$ is **coarser** than a partition $\mathscr{Q}'$ if every element of $\mathscr{Q}'$ is a subset of an element of $\mathscr{Q}$.

The link with measure theory is the following. To each partition $\mathscr{Q}$ of $\mathscr{C}$ we associate a σ-**algebra** $\mathscr{F}$ on $\mathscr{C}$ as follows: $\mathscr{F}$ is a subset of $2^{\mathscr{C}}$, and an element of $2^{\mathscr{C}}$ belongs to $\mathscr{F}$ if and only if it is a union (possibly empty) of elements of $\mathscr{Q}$. One checks immediately that $\mathscr{F}$ is indeed a σ-algebra, i.e. is stable under complementation and countable unions: if $A \in \mathscr{F}$ its complement $A^c \in \mathscr{F}$, and is $A_i \in \mathscr{F}$, $i \in I$ (a finite or countable set), then $\bigcup_{i \in I} A_i \in \mathscr{F}$. Note that one has stability under arbitrary unions in that case, but this is not required for a σ-algebra. The fact that the partition $\mathscr{Q}'$ is finer than $\mathscr{Q}$ reads as the inclusion $\mathscr{F} \subset \mathscr{F}'$ at the level of associated σ-algebras.

Which σ-algebras on $\mathscr{C}$ arise in this way? The answer is that if $\mathscr{C}$ is finite or countable, all of them do, and there is a one to one correspondence between partitions and σ-algebras. To go back from a σ-algebra $\mathscr{F}$ to a partition $\mathscr{Q}$ when $\mathscr{C}$ is finite or countable, define an **atom** of $\mathscr{F}$ to be a non-empty element of $\mathscr{F}$ which does not contain properly any other non-empty element of $\mathscr{F}$. Then *the atoms form a partition of $\mathscr{C}$* (you can try to prove it).

We now introduce an index t (belonging to some ordered set $(T, \leq)$, typically $T = \mathbb{N}_n = \{0, 1, \ldots, n\}$, $T = \mathbb{N} = \{0, 1, \ldots\}$ or $T = \mathbb{R}^+ = [0, +\infty[$) which will be identified with "time", and introduce a family $(\mathscr{Q}_t)_{t \in T}$ of partitions of $\mathscr{C}$, which get finer and finer as t increases. By convention $\mathscr{Q}_0$ is the trivial partition with $\mathscr{C}$ as its single piece.

The family of σ-algebras $(\mathscr{F}_t)_{t \in T}$ associated to the family of partitions $(\mathscr{Q}_t)_{t \in T}$ is called a **filtration** of $\mathscr{C}$: $\mathscr{F}_s \subset \mathscr{F}_t$ for $s \leq t$.

After this mathematical preliminary, we can turn to more familiar constructions by assuming that $\mathscr{C}$ is the configuration space of a statistical model. Very often one works with discrete variable in finite volume, and then $\mathscr{C}$ is discrete, and finite or countable. That's one of the reasons why physicists are not used to σ-algebras, which are the appropriate tool to deal with general configurations spaces, but which can be replaced by partitions as seen above for finite or countable configuration spaces. So we assume that $\mathscr{C}$ is finite or countable for the rest of this discussion.

Before we start, let us quote one famous instance where the above ideas are used implicitly all the time: the **renormalisation group**. If $\mathscr{C}$ describes the configurations of a spin system, one can partition $\mathscr{C}$ for instance by first partitioning "space" into blocks of spins, and then regrouping all configurations for which the magnetisation of each block is given, but forgetting the other details inside a block. Taking larger and larger blocks means taking coarser and coarser partitions leading to a filtration, and increasing t means keeping finer and finer details of small scales physics. The reader is invited to reinterpret the following constructions (conditional expectations, ...) in terms of familiar concepts (effective actions, ...).

Let W_c, $c \in \mathscr{C}$, be a family of Boltzmann weights on $\mathscr{C}$. We assume that $W_c \geq 0$ for each c and that $Z := \sum_c W_c \in]0, +\infty[$, so that $\mathbf{P}(c) = W_c/Z$ defines a probability on $\mathscr{C}$.

To start with, let O_c, $c \in \mathscr{C}$ be an observable (that is, a map from $\mathscr{C}$ to $\mathbb{R}$) and assume that $\frac{1}{Z}\sum_c W_c|O_c| < +\infty$ which ensures that $\mathbb{E}(O) := \frac{1}{Z}\sum_c W_c O_c$, the expectation of O with respect to $\mathbf{P}$, is well defined. In probability theory, we would say that O is a $\mathbf{P}$-integrable random variable.

If $\mathscr{D}$ is a subset of $c \in \mathscr{C}$ and $\mathbf{P}(\mathscr{D}) := \frac{1}{Z}\sum_{c \in \mathscr{D}} W_c > 0$, one can do statistical mechanics on $\mathscr{D}$: we define $Z_{\mathscr{D}} := \sum_{c \in \mathscr{D}} W_c = Z\mathbf{P}(\mathscr{D})$ and $\mathbb{E}_{\mathscr{D}}(O) := \frac{1}{Z_{\mathscr{D}}}\sum_{c \in \mathscr{D}} W_c O_c$. This is a conditional expectation, and it looks trivial to physicists because statistical mechanics uses all the time partition functions, i.e. non-normalised but finite measures to compute probabilities.

A bit more generally, if $\mathscr{Q}$ is a partition of $\mathscr{C}$, with associated σ-algebra $\mathscr{F}$, we can define a new observable, denoted by $\mathbb{E}(O|\mathscr{F})$ as follows. For each $c \in \mathscr{C}$ there is a single $\mathscr{D} \in \mathscr{C}$ such that $c \in \mathscr{D}$. If $Z_{\mathscr{D}} > 0$, then $\mathbb{E}(O|\mathscr{F})_c := \mathbb{E}_{\mathscr{D}}(O)$. If $Z_{\mathscr{D}} = 0$, give $\mathbb{E}(O|\mathscr{F})_c$ an arbitrary value (but the same for all $c \in \mathscr{D}$). So $\mathbb{E}(O|\mathscr{F})$ is still an observable on $\mathscr{C}$, but a coarse-grained one: its value on c depends only on the piece of the partition to which c belongs, and is equal to the average of O on this piece. Beware, the conditional expectation $\mathbb{E}(O|\mathscr{Q})$ is not a number. It is an observable, which moreover has a nice characterisation: it is constant on every piece of $\mathscr{Q}$ and if U is any (bounded, this is just to make sure that all formulæ are convergent) observable on $\mathscr{C}$ which is constant on every piece of $\mathscr{Q}$ then $\mathbb{E}(OU) = \mathbb{E}(\mathbb{E}(O|\mathscr{F})U)$. The reader should check that this is a characterisation. There is a small subtlety: $\mathbb{E}(O|\mathscr{F})$ is fully determined only on the pieces of $\mathscr{Q}$ with non-vanishing partition function; there can be several version of the conditional expectation, but they differ only on a sets with 0 **P**-measure.

Now we use a filtration and define $O(t) := \mathbb{E}(O|\mathscr{F}_t)$ for $t \in T$. Then, a trivial computation reveals that $\mathbb{E}(O(t)|\mathscr{F}_s) = O(s)$ for $s \le t \in T$, i.e. $(O(t))_{t \in T}$ is a martingale. In fact, it is a special case of martingale, a closed martingale, because there is an observable, O itself, such that $O(t) := \mathbb{E}(O|\mathscr{F}_t)$. This is the content of the sentence "Observables, or correlation functions, are martingales".

It is useful to go a bit further. Suppose that $\tilde{W}_c$, $c \in \mathscr{C}$ is another family of Boltzmann weights on $\mathscr{C}$, such that $\tilde{W}_c \ge 0$ for each c and that $\tilde{Z} := \sum_c \tilde{W}_c \in]0, +\infty[$. Let $\tilde{\mathbf{P}}$ and $\tilde{\mathbb{E}}$ be the corresponding probability measure and expectation.

Let $\mathscr{Q}$ be a partition of $\mathscr{C}$ and assume that for any piece $\mathscr{D}$ of $\mathscr{Q}$ either $Z_{\mathscr{D}} > 0$ or $\tilde{Z}_{\mathscr{D}} := \sum_{c \in \mathscr{D}} \tilde{W}_c = 0$. Then we can define an observable R on $\mathscr{C}$ as follows. For each $c \in \mathscr{C}$ there is a single $\mathscr{D} \in \mathscr{C}$ such that $c \in \mathscr{D}$. If $Z_{\mathscr{D}} > 0$ then $R_c = \frac{\tilde{Z}_{\mathscr{D}}}{Z_{\mathscr{D}}}$. If $Z_{\mathscr{D}} = 0$, give R_c an arbitrary value (but the same for all $c \in \mathscr{D}$). If U is any $\tilde{\mathbf{P}}$-integrable observable on $\mathscr{C}$ which is constant on every piece of $\mathscr{Q}$ then RU is a $\mathbf{P}$-integrable observable on $\mathscr{C}$ which is constant on every piece of $\mathscr{Q}$ and $\tilde{\mathbb{E}}(U) = \mathbb{E}(RU)$. This is expressed mathematically as follows (this is mostly vocabulary): when coarse-grained down to $\mathscr{Q}$, the measure $\tilde{\mathbf{P}}$ is absolutely continuous with respect to $\mathbf{P}$, and the Radon-Nykodim derivative $\frac{d\tilde{\mathbf{P}}}{d\mathbf{P}}$ is equal to R.

Now, the case of a filtration. We assume that for any $t \in T$ and any piece $\mathscr{D}$ of the partition $\mathscr{Q}_t$, either $Z_{\mathscr{D}} > 0$ or $\tilde{Z}_{\mathscr{D}} = 0$. We define $R(t)$ as before, as the ratio of partition functions for w and $\tilde{W}$ on any piece of the partition $\mathscr{Q}_t$. Then $R(t)$ is also the Radon-Nykodim derivative $\frac{d\tilde{\mathbf{P}}}{d\mathbf{P}}$ when both measures are coarse-grained down to $\mathscr{Q}_t$. The following computation will show that $\mathbb{E}(R(t)|\mathscr{F}_s) = R(s)$ for $s \le t \in T$, i.e. that $(R(t))_{t \in T}$ is a martingale. To show that $R(t)$ is a martingale, we use the above characterisation of conditional expectations.

We need to show that, if $t \ge s$ and if U is bounded and constant on every piece of $\mathscr{Q}_s$, $\mathbb{E}(U R(t)) = \mathbb{E}(U R(s))$. When X is an observable on $\mathscr{C}$ which is constant on a subset $\mathscr{D}$ of $\mathscr{C}$, we write $X_{\mathscr{D}}$ for the value of X_c when $c \in \mathscr{D}$. We compute Z (which is non-zero) times the right-hand side $\mathbb{E}(U R(s))$. This is

$$Z\mathbb{E}\big(U R(s)\big) = \sum_{c\in\mathscr{C}} U_c R(s)_c W_c$$

$$= \sum_{\mathscr{D}\in\mathscr{Q}_s} \sum_{c\in\mathscr{D}} U_c R(s)_c W_c$$

$$= \sum_{\mathscr{D}\in\mathscr{Q}_s} U_{\mathscr{D}} R(s)_{\mathscr{D}} \sum_{c\in\mathscr{D}} W_c$$

$$= \sum_{\mathscr{D}\in\mathscr{Q}_s} U_{\mathscr{D}} R(s)_{\mathscr{D}} Z_{\mathscr{D}}$$

$$= \sum_{\mathscr{D}\in\mathscr{Q}_s} U_{\mathscr{D}} \tilde{Z}_{\mathscr{D}}.$$

As $t \geq s$, U is a fortiori constant on every piece of $\mathscr{Q}_t$ and a parallel computation yields

$$Z\mathbb{E}\big(U R(t)\big) = \sum_{\mathscr{D}\in\mathscr{Q}_t} U_{\mathscr{D}} \tilde{Z}_{\mathscr{D}}.$$

The two expression are equal. Indeed, $\tilde{Z}_{\mathscr{D}}$ is an additive functional of $\mathscr{D}$: its value on a disjoint union is the sum of the values on each piece of the union. Use this to compute $\sum_{\mathscr{D}\in\mathscr{Q}_s} U_{\mathscr{D}} \tilde{Z}_{\mathscr{D}}$ by breaking each $\mathscr{D}\in\mathscr{Q}_s$ into its $\mathscr{Q}_t$ pieces.

Were the support of $\tilde{W}$ included in that of W (i.e. if $\tilde{W}_c$ vanished for all c's for which $W_c = 0$), we could define an observable O by $O_c = \tilde{W}_c / W_c$ whenever the denominator is non-zero and keep O_c arbitrary when $W_c = 0$. We would be back to the case of observables: $R(t)$ would simply equal $O(t)$ and be a closed martingale. But in general, $R(t)$ is not a closed martingale. However, we get the general principle "Ratios of partition functions are martingales".

The above considerations look a little bit like abstract nonsense. So let us apply them to interfaces (say on a finite lattice domain to keep a finite or countable configuration space). Suppose that to each configuration one can associate an interface, and $\mathscr{F}_t$ gives a finer description of this interface as t increases, for instance by describing completely larger and larger initial segments of the interface. Now one can be interested in certain configurations $c \in \mathscr{D}$ (resp. $\tilde{\mathscr{D}}$), which are specified by some feature of the interface they contain (for instance starting point, end point,...). Then we consider weights W_c (resp. $\tilde{W}_c$) which vanish outside $\mathscr{D}$ (resp. $\tilde{\mathscr{D}}$). If the possible initial segments of interfaces (at least for t up to a certain value) are the same for $\mathscr{D}$ and $\tilde{\mathscr{D}}$, we get martingales by taking ratios of partial partition functions as explained above.

Going to the continuum limit requires some care: the previous argument seems to be closely related to the existence of *Radon-Nykodim derivatives*, which can be tricky when taking the continuum limit. But when this is known, the fact that ratios of partition functions are martingales is quite a powerful tool.

For instance, in the computation of locality, we obtained stochastic differential equations allowing to compare chordal SLE from 0 to ∞ to chordal SLE from 0 to a point x (say $x > 0$) at finite distance. Obviously the second measure on the full set of curves cannot be absolutely continuous with respect to the first because the event to hit x has probability 0 for the first (at least if $\kappa < 8$) and 1 for the second. However, the two measures are absolutely continuous with respect to each other when coarse-grained to ignore what happens to a curve once it has hit the interval $]x, +\infty[$. We have already observed that the ratios of partition functions are (local) martingales in that case.

2.6 Notes and References

For a very short overview of SLE, see [25]. The lectures notes [8] concentrate on discrete models and Loewner chains. There are many reviews on SLE. Historically, the first one is [27]. The reviews [18, 24] are a valuable source of information for interfaces and SLE, but mostly seen from the CFT side. Reference [9] is a review on growth processes, Loewner chains, SLE and CFT written by physicists, but with a strong emphasis on probabilistic aspects.

Let us note that Loewner chains for certain models, for instance Laplacian growth, are related to integrable systems, see e.g. [8, 9] and references therein.

The interpretation of extended objects in $2D$ critical phenomena in terms of logarithmic CFT is only starting, but some interesting approaches and results can be found in [41] for crossing probabilities and [30] for SLE martingales.

The proof of convergence of the exploration process to a conformally invariant continuum limit is announced in [47] and proved along rather different lines in [15, 16].

The proof that the off-critical continuum percolation measure is singular with respect to the critical continuum percolation measure can be found in [42].

The canonical self-avoiding walk puts the uniform measure on all lattice paths of a given length. This model has proved remarkably difficult to tackle. Loop-erased random walks are introduced in [32] as an example of walks which are self avoiding but more tractable than the canonical self avoiding walk.

The seminal paper on SLE is [45], which dealt mainly with loop-erased random walks. The existence a conformal invariance of a continuum limit of the loop-erased random walk measure is proved in [38, 39].

The Riemann mapping theorem is proved in a number of textbooks, see e.g. [19]. Most proofs involve a good amount of functional analysis, related for instance to the existence of the Green function of the Laplace operator with Dirichlet boundary conditions. The deep relationship between the Laplace operator and Brownian motion can be used to give a probabilistic proof of Riemann's theorem, see e.g. [1].

The seminal paper on Loewner chains is [40]. A mathematical introduction to Loewner chains can be found in [19], or with more details in [43].

Among the physicists works that inspired mathematical work, culminating in SLE, one should quote Cardy's percolation formula [17], and intersection exponents [21]. Though physicists understood quite well conformal invariance for correlation functions of local observables since 1984 [14], they had failed to find the axiomatic framework describing conformally invariant measures on interfaces.

After the seminal millennium paper by Schramm [45], Lawler, Schramm and Werner embarked in an impressive series of contributions where they explored SLE (locality, restriction, ...) but also its applications to a number of open problem like intersection exponents, the dimension of the Brownian frontier, ... see e.g. [33–39].

The proof that the SLE trace is a curve is surprisingly deep, especially at $\kappa = 8$, see [44].

The proof that the Hausdorff dimension of SLE_κ is $\min(1 + \kappa/8, 2)$ is also very tricky. The first proof appeared in [13].

The proof of duality of chordal SLE is in [48] and that of reversibility is in [49], note also the nice (but inconclusive) approach in [31].

The seminal article on the SLE-CFT correspondence is [2]. Reference [22] deals with $\kappa = 8/3$ i.e. the case when the SLE hull is a simple curve and the conformal anomaly vanishes ($c = 0$).

The computation of the hitting probability (and others) via CFT techniques is explained in [3]. The relations between SLE martingales and the representation theory of the Virasoro algebra is explained in [4, 6] and generalised in [29] (see also [30] for applications to logarithmic CFT). Locality via CFT and partition functions is explained in [6].

A better understanding of why SLE and CFT are related the way they are came via the double counting argument explained in Sect. 2.5.5 in these notes. The case of observables is explained in [10]. More games with partition functions, conditioning and Girsanov's theorem can be found in [12, 26, 46].

The relationship between other kinds of SLE's (a subject not pursued in these notes) and CFT is explained in [5, 7, 10].

Multiple interfaces (corresponding to an arbitrary number of changes in boundary conditions) are described in [11, 20]. The first one insists a constraint called commutation, while the second exploits partition function techniques. See also [23] for a nice interpretation.

A very readable introduction to stochastic calculus is [28].

References

1. Bass, R.F.: Probabilistic Techniques in Analysis. Springer, New York (1995)
2. Bauer, M., Bernard, D.: SLE_κ growth processes and conformal field theories. Phys. Lett. B **543**, 135 (2002)
3. Bauer, M., Bernard, D.: Conformal field theories of stochastic Loewner evolutions. Commun. Math. Phys. **239**, 493 (2003)
4. Bauer, M., Bernard, D.: SLE martingales and the Virasoro algebra. Phys. Lett. B **557**, 309 (2003)
5. Bauer, M., Bernard, D.: SLE, CFT and zig-zag probabilities. In: Lawler, G., Khanin, K., Norris, J. (eds.) Conformal Invariance and Random Spatial Processes. NATO Advanced Study Institute. NATO, Brussels (2003)
6. Bauer, M., Bernard, D.: Conformal transformations and the SLE partition function martingale. Ann. Henri Poincaré **5**, 289 (2004)
7. Bauer, M., Bernard, D.: CFTs of SLEs: the radial case. Phys. Lett. B **583**, 324 (2004)

8. Bauer, M., Bernard, D.: Loewner chains. In: Baulie, L., et al. (eds.) String Theory: From Gauge Interactions to Cosmology. NATO Advanced Study Institute. Springer, Dordrecht (2004)
9. Bauer, M., Bernard, D.: $2D$ growth processes: SLE and Loewner chains. Phys. Rep. **432**, 115 (2006)
10. Bauer, M., Bernard, D., Houdayer, J.: Dipolar SLEs. J. Stat. Mech. P03001 (2005)
11. Bauer, M., Bernard, D., Kytölä, K.: Multiple Schramm-Loewner evolutions and statistical mechanics martingales. J. Stat. Phys. **120**, 1125 (2005)
12. Bauer, M., Bernard, D., Kennedy, T.-G.: Conditioning SLEs and loop-erased random walks. J. Math. Phys. **50**, 043301 (2009)
13. Beffara, V.: The dimension of the SLE curve. Ann. Probab. **36**, 1421 (2008)
14. Belavin, A.A., Polyakov, A.M., Zamolodchikov, A.B.: Infinite conformal symmetry in two-dimensional quantum field-theory. Nucl. Phys. B **241**, 333 (1984)
15. Camia, F., Newman, C.M.: Two-dimensional critical percolation: the full scaling limit. Commun. Math. Phys. **268**, 1 (2006)
16. Camia, F., Newman, C.M.: Critical percolation exploration path and SLE$_6$: a proof of convergence. Probab. Theory Relat. Fields **139**, 473 (2007)
17. Cardy, J.L.: Critical percolation in finite geometries. J. Phys. A, Math. Gen. **25**, 201 (1992)
18. Cardy, J.L.: SLE for physicists. Ann. Phys. **318**, 81 (2005)
19. Conway, J.H.: Functions of One Complex Variable I and II. Springer, Heidelberg (1995)
20. Dubédat, J.: Commutation relations for SLE. Commun. Pure Appl. Math. **60**, 1792 (2007)
21. Duplantier, B., Kwon, K.-H.: Conformal invariance and intersection of random walks. Phys. Rev. Lett. **61**, 2514 (1988)
22. Friedrich, R., Werner, W.: Conformal restriction, highest weight representations and SLE. Commun. Math. Phys. **243**, 105 (2003)
23. Graham, K.: On multiple Schramm-Loewner evolutions. J. Stat. Mech. P03008 (2007)
24. Gruzberg, I.: Stochastic geometry of critical curves, Schramm-Loewner evolutions, and conformal field-theory. J. Phys. A, Math. Gen. **39**, 12601 (2006)
25. Gruzberg, I., Kadanov, L.: The Loewner equation: maps and shapes. J. Stat. Phys. **114**, 1183 (2004)
26. Hagendorf, C., Bernard, D., Bauer, M.: The Gaussian free field and SLE$_4$ on doubly-connected domains. J. Stat. Phys. **140**, 1 (2010)
27. Kager, W., Nienhuis, B.: A guide to stochastic Loewner evolution and its applications. J. Stat. Phys. **15**, 1149 (2004)
28. Kuo, H.-H.: Introduction to Stochastic Integration. Springer, Heidelberg (2006)
29. Kytölä, K.: Virasoro module structure of local martingales of SLE variants. Rev. Math. Phys. **19**, 455 (2007)
30. Kytölä, K.: SLE local martingales in logarithmic representations. J. Stat. Mech. P08005 (2009)
31. Kytölä, K., Kemppainen, A.: SLE local martingales, reversibility and duality. J. Phys. A, Math. Gen. **39**, 657 (2006)
32. Lawler, G.F.: A self-avoiding walk. Duke Math. J. **47**, 655 (1980)
33. Lawler, G.F., Schramm, O., Werner, W.: Values of Brownian intersections exponents I: half-plane exponents. Acta Math. **187**, 237 (2001)
34. Lawler, G.F., Schramm, O., Werner, W.: Values of Brownian intersections exponents II: plane exponents. Acta Math. **187**, 275 (2001)
35. Lawler, G.F., Schramm, O., Werner, W.: The dimension of the planar Brownian frontier is $4/3$. Math. Res. Lett. **8**, 401 (2001)
36. Lawler, G.F., Schramm, O., Werner, W.: One-arm exponent for $2D$ critical percolation. Electron. J. Probab. **7**, 2 (2001)
37. Lawler, G.F., Schramm, O., Werner, W.: Values of Brownian intersections exponents III: two-sided exponents. Ann. Inst. Henri Poincaré **38**, 109 (2002)
38. Lawler, G.F., Schramm, O., Werner, W.: Conformal invariance of planar loop-erased random walks and uniform spanning tress. Ann. Probab. **32**, 939 (2004)

39. Lawler, G.F., Schramm, O., Werner, W.: On the scaling limit of planar self-avoiding walk. Proc. Symp. Pure Math. **72**, 339 (2004)
40. Loewner, K.: Untersuchungen über schlichte konforme Abbildungen des Einheitskreises. Math. Ann. **89**, 103 (1923)
41. Mathieu, P., Ridout, D.: From percolation to logarithmic conformal field theory. Phys. Lett. B **657**, 120 (2007)
42. Nolin, P., Werner, W.: Asymmetry of near-critical percolation interfaces. J. Am. Math. Soc. **22**, 797 (2009)
43. Pommerenke, C.: Univalent Functions. With a Chapter on Quadratic Differentials by Gerd Jensen. Vandenhoeck & Ruprecht, Göttingen (1975)
44. Rohde, S., Schramm, O.: Basic properties of SLE. Ann. Math. **161**, 879 (2005)
45. Schramm, O.: Scaling limits of loop-erased random walks and uniform spanning trees. Isr. J. Math. **118**, 221 (2000)
46. Schramm, O., Wilson, D.: SLE coordinate change. N.Y. J. Math. **11**, 659 (2005)
47. Smirnov, S.: Critical percolation in the plane: conformal invariance, Cardy's formula, scaling limits. C. R. Acad. Sci. Paris **333**, 239 (2001)
48. Zhan, D.: Duality of chordal SLE. Invent. Math. **174**, 309 (2008)
49. Zhan, D.: Reversibility of chordal SLE. Ann. Probab. **36**, 1472 (2008)

Chapter 3
Numerical Tests of Schramm-Loewner Evolution in Random Lattice Spin Models

Christophe Chatelain

3.1 Introduction

Schramm-Loewner Evolution (or Stochastic Loewner Evolution (SLE)) allows for the construction of a family of non-intersecting and non-branching continuous two-dimensional random curves by means of a local Markovian growth process. These curves, called **SLE traces**, are *fractal* and their probability distribution is invariant under conformal transformation. (Chordal) SLE depends on a single real positive parameter denoted κ. By extension, the different families of random curves will be denoted SLE_κ. The continuum limit of some simple generalisations of the random walk on a regular two-dimensional lattice is believed to be SLE traces. The paradigmatic example is the **loop-erased random walk** (LERW) which is presented in Fig. 3.1 and for which it has been shown that $\kappa = 2$ [34, 48]. One should also mention percolation clusters whose boundaries are SLE_6 traces [9, 45, 51] or the Self-Avoiding Walk (SAW) that has been conjectured to be $\text{SLE}_{8/3}$ [35]. These examples provide good reasons to believe that SLE traces can also be found in the usual spin models of Statistical Physics, like the celebrated Ising [28], Potts [43], or $O(n)$ models. These models have in common to possess a representation in terms of loops. Moreover, the Potts model reduces to percolation in the limit $q \to 1$ and the $O(n)$ model to SAW when $n \to 0$ and to LERW for $n = -2$. Furthermore, conformal invariance holds for an infinite set of two-dimensional models, including Ising and Potts models. Since at criticality spin clusters are fractal, with fractal dimension related to critical exponents [16, 53], one may infer that SLE traces might be recovered as the continuum limit of interfaces between spin clusters. SLE would thus offer a new description of these critical systems leading to new exact results about models of Statistical Physics. For the pure Ising model, a rigorous proof that critical curves are SLE traces with $\kappa = 3$ has been given by Smirnov [52]. However,

C. Chatelain (✉)
GPS—DP2M—IJL (CNRS UMR 7198), Université de Lorraine Nancy, BP 70239, 54506 Vandœuvre-lès-Nancy Cedex, France
e-mail: Christophe.Chatelain@ijl.nancy-universite.fr

M. Henkel, D. Karevski (eds.), *Conformal Invariance: an Introduction to Loops, Interfaces and Stochastic Loewner Evolution*, Lecture Notes in Physics 853, DOI 10.1007/978-3-642-27934-8_3, © Springer-Verlag Berlin Heidelberg 2012

Fig. 3.1 The *black curve* is an example of a non-intersecting and non-branching continuous curve obtained with the following algorithm: a random walk is generated and each time the curve touches itself, the loop that has been formed is erased. In *grey* are plotted the lattice sites visited by the curve but then erased because they were belonging to a loop. This model is known as the loop-erased random walk and has been shown to be described by SLE$_2$

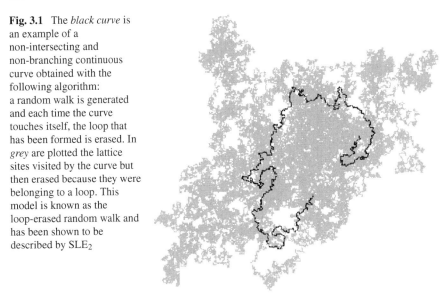

for most of the models, no exact result is available and the best estimates of critical exponents were provided by simulations. Numerical techniques are often the only possibility to check whether a given critical curve may be described by SLE or not. It is especially the case for random systems.

The goal of this review is to present numerical methods and results for lattice spin models in the perspective of checking the validity of SLE. We shall first discuss the genesis of an interface in a critical lattice spin model and the algorithms for its localisation. Different tests of SLE are then presented in the second section. To illustrate the text, original data for the celebrated **Ising model** defined by the hamiltonian

$$\mathcal{H} = -J \sum_{(i,j)} \sigma_i \sigma_j, \quad \sigma_i \in \{+1, -1\} \tag{3.1}$$

(with the usual nearest-neighbour interactions) and its generalisation, the q-state **Potts model**

$$\mathcal{H} = -J \sum_{(i,j)} \delta_{\sigma_i, \sigma_j}, \quad \sigma_i = 0, \ldots, q - 1 \tag{3.2}$$

will be presented. It is not our aim here to give an exhaustive list of results for pure models. Finally, tests of a possible description of interfaces in random systems by SLE are presented and discussed in the last section.

3.2 Genesis and Localisation of the Interface

At the critical point, clusters of different spin states appear spontaneously and the interfaces between them are fractal. However, most of them surround small clusters

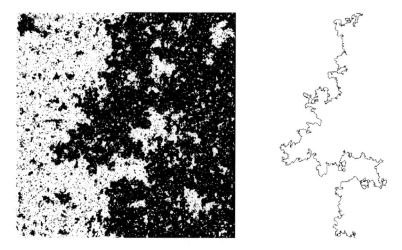

Fig. 3.2 On the *left*, example of an interface between two clusters of different spin states in the Ising model. *Each pixel* is a spin and its *colour* (*black* or *white*) corresponds to one of the two possible states. The interface is plotted in *blue*. Its existence is ensured by the boundary conditions. On the *right*, only the interface is plotted

and therefore form closed loops unsuitable for a description by SLE. In the limit of an infinite system, one cluster is expected to **percolate**, i.e. to span the whole lattice. From the frontiers of this cluster, one could in principle determine curves joining two points on two different boundaries of the system. In numerical simulations, we have to face an additional problem: because of the finite size of the lattice, the **percolating cluster** may not be unique or may not exist at all! To ensure the existence of a percolating cluster, a convenient way is to choose symmetry-breaking boundary conditions. In the case of the Ising model for example, one can fix the spins to the state $+1$ on one half of the boundary and -1 on the other half. The **interface** is the continuous curve on the dual lattice that starts at the boundary and for which the two spins at both sides have different states. An example of such an interface is given in Fig. 3.2. It is important to note that closed loops around spin clusters are not taken into account. Because of the construction of the interface, it necessarily ends at the boundary. As a consequence, the SLE that might be obtained in the continuum limit is the so-called chordal-SLE for which both starting and ending points are on the boundary $\partial \mathbb{D}$ of the domain $\mathbb{D}$.

We shall now discuss the localisation of the interface for a given spin configuration. Spins σ_i are placed on the nodes i of the lattice. When two neighbouring spins, say σ_i and σ_j are in the same state, we shall say that a bond is put on the lattice link joining the two sites i and j. When $\sigma_i \neq \sigma_j$, the link remains empty. Since it is always perpendicular to lattice links, the interface lies on the so-called dual lattice. An example of such a dual lattice is shown as dashed lines in Fig. 3.3. By construction, spins on both sides of the interface are in different states. It means that the interface crosses only empty links. The algorithm for the construction of the interface could be the following: start from one of the two empty links on the boundary of the lattice. The tip of the interface moves along the link of the dual lattice perpendicular

Fig. 3.3 Construction of the
interface on the hexagonal
lattice. The links of the
hexagonal lattice are depicted
as *continuous lines*. Spins are
the *filled circles* on the nodes
of the lattice. The *two
colours*, *white* or *black* of
these *circles*, correspond to
the two possible spin states of
the Ising model (± 1). The
links of the dual lattice are
represented as *dashed lines*.
The interface is the *bold red
line*

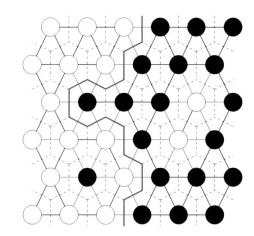

to it. Arriving at the next node of the dual lattice, consider the set of links on which
the interface could continue its movement. Note that it cannot go backwards. On the
hexagonal lattice, the dual lattice is triangular so there are only two possible movements left for the tip. For each one of these links of the dual lattice, consider the
spins on both sides of them. The tip has to choose the link for which the spins on
both sides are in different states. As can be seen in Fig. 3.3, only one of the possible links satisfies this condition. The interface is thus unambiguously defined. The
operation is repeated until the tip of the interface reaches the other boundary.

As mentioned above, the interface is unambiguously defined. It is not necessarily the case for another lattice. It is in particular not the case for the square lattice.
As can be shown in Fig. 3.4, the dual lattice is a square lattice, too. It means that
the tip of the interface has now to choose in which direction to move among three
possibilities. For most of the spin configurations, only one direction points towards
an empty link of the lattice. It means that there are two bonds on the plaquette and
thus two empty links allowing the tip to enter and exit the plaquette. However, one
can imagine spin configurations for which there are four empty links. Consider for
example the spin configuration of Fig. 3.5. The tip is allowed to take any of the
three remaining directions but we have specified no rule for it to choose one of
them. The simplest rule one could imagine is to allow the tip to choose randomly
one of the three possible directions. It turns out that this rule introduces a systematic deviation in the fractal dimension of the interface [47]. Another rule has been
thus suggested: when the tip encounters a plaquette allowing for several possible
movements, it chooses always to turn on the left (or the right). This rule is called
left tie-breaking algorithm (or right tie-breaking algorithm). It has been checked
that the fractal dimension is correctly recovered with this algorithm [47]. However,
it does not guaranty that other quantities, left-passage probability for example, are
not affected by this choice The reader may wonder why we insist in using a square
lattice instead of a hexagonal one. The answer is that the square lattice is self-dual
while the hexagonal one is not (its dual is the triangular lattice). As a consequence,
the critical point is often known exactly on a square lattice but more rarely on the
hexagonal lattice.

Fig. 3.4 Construction of the
interface on the square lattice.
The same conventions as in
Fig. 3.3 have been used for
links of the lattice and the
dual lattice and spins. The
interface is again the *bold red
line*. As explained in the text,
it is not unambiguously
defined: the *dashed loop*
could be considered part of
the interface

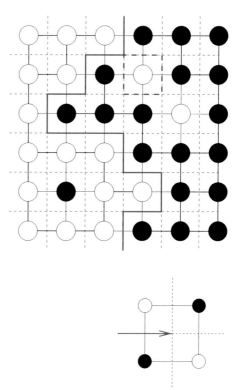

Fig. 3.5   A problematic spin
configuration on a square
plaquette for the construction
of the interface

To circumvent the difficulties related to the construction of the interface on the
square lattice, one can define it on the so-called **medial lattice** rather than on the
dual one. The medial lattice is a square lattice but rotated by 45° from the original
one. Its nodes lie at the centre of the links of the square lattice. Spins are at the
centre of plaquettes of the medial lattice. Only half of the plaquettes are occupied
by a spin. A direction is assigned to each link of the medial lattice. By convention,
one can choose to turn clockwise around each plaquette occupied by a spin. An
example of medial lattice is shown in Fig. 3.6. The algorithm for the construction
of the interface remains the same: the tip jumps from one node of the medial lattice
to the other one by following the links whose direction points outward. It can be
seen in Fig. 3.6 that at each nodes of the medial lattice, two links go inwards and
the two others outward. As a consequence, a tip has always only two possibilities
to leave a node. Its choice is made according to the spins on both sides of these two
links or equivalently by forbidding the tip to cross a bond of the original lattice. The
interface is unambiguously defined on the medial lattice.

Up to now, we have only considered the example of the Ising model. An additional problem arises in models where spins are allowed to take more than two
states, typically for the Potts model with $q > 2$. To induce an interface, one may
use the same conditions as for the Ising model: spins are fixed in the state $\sigma = 0$ on
one half of the boundary and $\sigma = 1$ in the other half. It means for the q-state Potts

Fig. 3.6 Construction of the
interface on the medial lattice

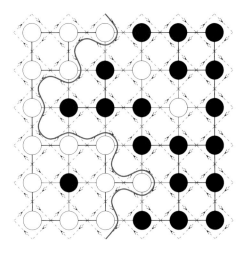

model that the $q - 2$ states $\sigma = 2, 3, \ldots, q - 1$ are not allowed on the boundary.
These boundary conditions will favour a large cluster of spins in the state $\sigma = 0$, for
example on the left of the lattice and a second one in the state $\sigma = 1$ on the right.
However, other clusters in states $\sigma = 2, 3, \ldots, q - 1$ may appear between these two
large clusters favoured by the boundary conditions. The interface cannot be defined
as the continuous curve with spins $\sigma = 0$ on the left and $\sigma = 1$ on the right. One may
use a less restrictive definition of the interface: for example, spins has chosen to be
in the state $\sigma = 0$ on the left but can be in any other state on the right of the interface.
In a sense, the problem has been mapped to an Ising model since from the point of
view of the interface, spins are in the state $\sigma = 0$ or $\sigma \neq 0$. As discussed by Gamsa
and Cardy [24], the symmetry between the $q - 1$ states $\sigma \neq 0$ must be preserved.
The boundary conditions should be defined in the following way: on one half of the
boundary, spins are fixed in the state $\sigma = 0$ while on the other half, they are free to
flip during the Monte Carlo simulations as long as they do not take the state $\sigma = 0$.
These boundary conditions were called **Fluctuating Boundary Conditions**. They
are equivalent to an infinite magnetic field $+h$ on one half of the boundary and $-h$
on the other half with the interaction hamiltonian $-h \sum_i \delta_{\sigma_i,0}$. The mapping to an
Ising model is then complete and the interface is unambiguously defined on the dual
of the hexagonal lattice or on the medial lattice. An example for the 3-state Potts
model is presented in Fig. 3.7.

In the Potts model, interfaces may be defined not only between clusters in different spin states but also between **Fortuin-Kasteleyn** (FK) **clusters** [23]. These
clusters arise from a representation of the partition function in terms of bond configurations instead of spin configurations. Loops surrounding these clusters on the
medial lattice appear naturally and play a central role in the equivalence of the Potts
model with a Coulomb gas. While interfaces between spin clusters suffer from difficulties, this is not the case for the interfaces between FK clusters. We assign a
bond variable $b_{ij} \in \{0; 1\}$ to each pair of neighbouring sites i and j. The partition

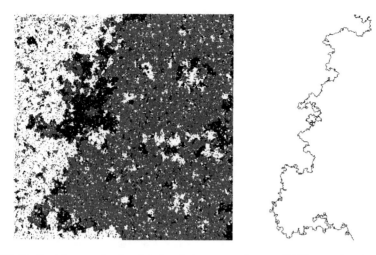

Fig. 3.7 On the *left*, example of an interface between two clusters of different spin states in the 3-state Potts model with fluctuating boundary conditions. *Each pixel* is a spin and its *colour* (*black*, *white* or *red*) corresponds to one of the three possible states. The interface is plotted in *blue*. On the *right*, only the interface is plotted

function may be written

$$Z = \sum_{\{\sigma\}} e^{J \sum_{(i,j)} \delta_{\sigma_i,\sigma_j}}$$

$$= \sum_{\{\sigma\}} \prod_{(i,j)} \left[(e^J - 1)\delta_{\sigma_i,\sigma_j} + 1 \right]$$

$$= \sum_{\{\sigma\}} \prod_{(i,j)} \left[\sum_{b_{ij}=0}^{1} (e^J - 1)\delta_{\sigma_i,\sigma_j}\delta_{b_{ij},1} + \delta_{b_{ij},0} \right] \tag{3.3}$$

An FK bond $b_{ij} = 1$ freezes the relative state of the two spins σ_i and σ_j. In contradistinction with the previously defined bonds between spins, the absence of FK bond, i.e. $b_{ij} = 0$, does not put any constrain on σ_i and σ_j. The sum over the spin configurations can now be performed:

$$Z = \sum_{\{b_{ij}\}} (e^J - 1)^{\sum_{ij} b_{ij}} \sum_{\{\sigma_i\}} \prod_{(i,j)} \delta_{\sigma_i,\sigma_j}\delta_{b_{ij},1}$$

$$= \sum_{G=\{b_{ij}\}} (e^J - 1)^{b(G)} q^{C(G)} \tag{3.4}$$

where $b(G) = \sum_{ij} b_{ij}$ is the total number of (frozen) bonds of the bond configuration $G = \{b_{ij}\}$ (that we shall call graph in the following) and $C(G)$ is the number of independent clusters. This formulation allows for the generalisation of the Potts

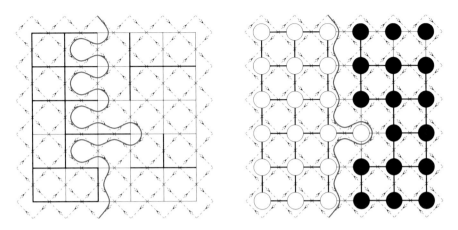

Fig. 3.8 On the *left*, example of an FK configuration. The bonds are represented as *thick lines* between nodes. The interface lives on the medial lattice and cannot cross these bonds. On the *right*, the fjords have been removed by first filling the FK cluster attached to the *left part of the boundary*. The sites belonging to the cluster are depicted by a *white circle* while the others carry a *black circle*. The same algorithm as for spin clusters is used to identify the interface, the so-called external perimeter of the FK cluster

model to non-integer values of the number of states q. Setting $p = 1 - e^{-J}$, we emphasise the relation between this representation and percolation:

$$Z = (1 - p)^{-N} \sum_G p^{b(G)} (1 - p)^{N - b(G)} q^{C(G)} \tag{3.5}$$

where N is the total number of links of the lattice. Percolation is recovered in the limit $q \to 1$. Both Swendsen-Wang [56] and Wolff [58] *cluster algorithms* rely on this representation of the partition function to accelerate drastically the Monte Carlo dynamics. From a spin configuration $\{\sigma\}$, a bond configuration is easily obtained by setting $b_{ij} = 1$ with probability $p = 1 - e^{-J}$ if $\sigma_i = \sigma_j$ and $b_{ij} = 0$ in any other case. On the boundaries where the spins are fixed, the FK bonds are frozen to the value $b_{ij} = 1$. A large FK cluster spans the lattice and it allows for the definition of an interface. Since the algorithm presented above was constructed upon bond variables (with the different definition $b_{ij} = 1$ if $\sigma_i = \sigma_j$), no modification is necessary to find the interface between FK clusters. The same subroutine may be used but supplied with different bond variables for interfaces between spin and FK clusters. Finally, it should be mentioned that FK clusters usually present **fjords**. Removing these fjords give access to the so-called external perimeter of the cluster. To do this, one can first recursively fill the FK cluster attached to the left boundary (in the context of cluster algorithms, the cluster is said to be decorated) and then define bond variables in the same way than between spins. Figure 3.8 gives an example of external perimeter.

3.3 Numerical Tests of SLE Properties

3.3.1 Domain Markov Property

In the following, we shall denote a **curve** γ in a **domain** $\mathbb{D}$ of the complex plane $\mathbb{C}$ joining the points a and b as

$$\gamma_{[ab]} : t \in [0; +\infty) \subset \mathbb{R} \longrightarrow \gamma_{[ab]}(t) \in \mathbb{D} \subset \mathbb{C} \qquad (3.6)$$

with $\gamma(0) = a$ and $\gamma(\infty) = b$. As mentioned in the introduction, the case of **chordal SLE** that we are interested in corresponds to $a, b \in \partial \mathbb{D}$. The curve being random, a probability distribution $P_{\mathbb{D}}(\gamma_{[ab]})$ of a particular realisation $\gamma_{[ab]}(t)$ can be *a priori* defined. The curve inherits two main properties from SLE: **conformal invariance** and **Domain Markovian Property**. The former will be discussed in the next section. The latter means that *for any curve $\gamma_{[ac]}(t)$ in a domain $\mathbb{D}$, joining the points $\gamma(0) = a$ and $\gamma(1) = b$ and passing through a third point $\gamma(t_0) = c$ with $t_0 \in [0; 1]$, the conditional probability of the second part $\gamma_{[cb]}$ of the curve is identical to the probability of the same curve in a domain where a cut has been made along the first part $\gamma_{[ac]}$:*

$$P_{\mathbb{D}}(\gamma_{[cb]} | \gamma_{[ac]}) = P_{\mathbb{D} \backslash \gamma_{[ac]}}(\gamma_{[cb]}) \qquad (3.7)$$

This is the **domain Markov property**, which is believed to hold true for a lattice spin model with short-range interactions. The argument is the following. On a lattice, the interface is a set of links of the dual lattice. The probability of an interface configuration is the sum of the Boltzmann weight over all possible spin configurations compatible with this interface. To be more precise, we shall consider the conditional probability $P_{\mathbb{D}}(\gamma_{[ab]} | \{\sigma\})$ of the interface $\gamma_{[ab]}$ given that the system is in the spin configuration $\{\sigma\}$. $P_{\mathbb{D}}(\gamma_{[ab]} | \{\sigma\})$ takes the value 1 if the spin configuration $\{\sigma\}$ induces the interface $\gamma_{[ab]}$ and 0 otherwise. The probability of $\gamma_{[ab]}$ is thus

$$P_{\mathbb{D}}(\gamma_{[ab]}) = \sum_{\{\sigma_i\}_{i \in \mathbb{D}}} P_{\mathbb{D}}(\gamma_{[ab]} | \{\sigma\}) P_{\mathbb{D}}(\{\sigma\}) = \sum_{\{\sigma_i\}_{i \in \mathbb{D}}} P_{\mathbb{D}}(\gamma_{[ab]} | \{\sigma\}) \frac{e^{-\mathcal{H}(\{\sigma\})/T}}{Z_{\mathbb{D}}}$$

$$(3.8)$$

Note that we assume that the spins on the boundary of the domain $\mathbb{D}$ are fixed in such a way that an interface is generated from a to b. The conditional probability follows

$$P_{\mathbb{D}}(\gamma_{[cb]} | \gamma_{[ac]}) = \frac{P_{\mathbb{D}}(\gamma_{[ac]} \cup \gamma_{[cb]})}{P_{\mathbb{D}}(\gamma_{[ac]})} = \frac{\sum_{\{\sigma_i\}_{i \in \mathbb{D}}} P_{\mathbb{D}}(\gamma_{[ac]} \cup \gamma_{[cb]} | \{\sigma\}) e^{-\mathcal{H}(\{\sigma\})/T}}{\sum_{\{\sigma_i\}_{i \in \mathbb{D}}} P_{\mathbb{D}}(\gamma_{[ac]} | \{\sigma\}) e^{-\mathcal{H}(\{\sigma\})/T}}$$

$$(3.9)$$

Performing a cut along a given path, say $\gamma_{[ac]}$, on the dual lattice is equivalent to remove the links that are perpendicular to this path and to fix the spins on both sides of the path. Accordingly to the symmetry-breaking boundary conditions chosen to generate the interface, the spins are fixed to different states on the left and on the right of the path. The Boltzmann weight on the domain $\mathbb{D}$ can be recovered simply by reintroducing the weight of the missing links $(i, j) \perp \gamma_{ac}$. In the case of the Ising model, it reads

$$e^{-\mathcal{H}/T} = Z_{\mathbb{D}} P_{\mathbb{D}}(\{\sigma\}) = Z_{\mathbb{D}\setminus\gamma_{[ac]}} e^{J\sum_{(i,j)\perp\gamma_{[ac]}}\sigma_i\sigma_j} P_{\mathbb{D}\setminus\gamma_{[ac]}}(\{\sigma\}) \qquad (3.10)$$

and the conditional probability leads to

$$P_{\mathbb{D}}(\gamma_{[cb]}|\gamma_{[ac]}) = \frac{\sum_{\{\sigma_i\}_{i\in\mathbb{D}}} P_{\mathbb{D}}(\gamma_{[ac]} \cup \gamma_{[ac]}|\{\sigma\}) e^{J\sum_{(i,j)\perp\gamma_{[ac]}}\sigma_i\sigma_j} P_{\mathbb{D}\setminus\gamma_{[ac]}}(\{\sigma\})}{\sum_{\{\sigma_i\}_{i\in\mathbb{D}}} P_{\mathbb{D}}(\gamma_{[ac]}|\{\sigma\}) e^{J\sum_{(i,j)\perp\gamma_{[ac]}}\sigma_i\sigma_j} P_{\mathbb{D}\setminus\gamma_{[ac]}}(\{\sigma\})}$$

$$(3.11)$$

The conditional probabilities $P_{\mathbb{D}}(\gamma_{[ac]} \cup \gamma_{[cb]}|\{\sigma\})$ and $P_{\mathbb{D}}(\gamma_{[ac]}|\{\sigma\})$ vanish unless the spins take different values on both sides of the path $\gamma_{[ac]}$. It implies that the sum over all spin configurations $\{\sigma_i\}_{i\in\mathbb{D}}$ can be restricted to those for which these spins have been transferred to the boundary, i.e. $\{\sigma_i\}_{i\in\mathbb{D}\setminus\gamma_{[ac]}}$. It remains

$$P_{\mathbb{D}}(\gamma_{[cb]}|\gamma_{[ac]}) = \frac{e^{-J\sum_{(i,j)\perp\gamma_{[ac]}} 1} \sum_{\{\sigma_i\}_{i\in\mathbb{D}\setminus\gamma_{[ac]}}} P_{\mathbb{D}}(\gamma_{[ac]} \cup \gamma_{[cb]}|\{\sigma\}) P_{\mathbb{D}\setminus\gamma_{[ac]}}(\{\sigma\})}{e^{-J\sum_{(i,j)\perp\gamma_{[ac]}} 1} \sum_{\{\sigma_i\}_{i\in\mathbb{D}\setminus\gamma_{[ac]}}} P_{\mathbb{D}}(\gamma_{[ac]}|\{\sigma\}) P_{\mathbb{D}\setminus\gamma_{[ac]}}(\{\sigma\})}$$

$$(3.12)$$

The exponential prefactors cancel and the denominator is equal to $P_{\mathbb{D}\setminus\gamma_{[ac]}}(\gamma_{[ac]}) = 1$ because the interface $\gamma_{[ac]}$ always exist in the domain $\mathbb{D} \setminus \gamma_{[ac]}$. We finally obtain the Domain Markov Property (3.7).

The derivation depends crucially on the fact that the interactions are local. Moreover, it *cannot* be generalised to systems with quenched randomness. In the example of the random-bond Ising model, Eq. (3.10) does not hold anymore because the coupling constant J is then site-dependent and the Boltzmann weight needs to be averaged over randomness. For both long-range interactions and random systems, it may be interesting to test the Domain Markov Property (3.7). However, the difficulty is that the dimension of the configuration space of the interface is extremely large. It would be much too memory-consuming to try to construct a histogram of all possible configurations during a Monte Carlo simulation. Partial tests are nevertheless possible, even though difficult. As far as we know, the Markovian property has been tested only for Ising Spin Glasses [8] and the Random Field Ising Model [54]. These tests will be discussed in the last section.

3.3.2 Conformal Invariance

The second property of SLE is **conformal invariance**: *the image under a conformal transformation $z \to f(z)$ of any curve $\gamma_{[ab]}$ in a domain $\mathbb{D}$ is a curve $(f \circ \gamma)$ in $f(\mathbb{D})$ joining the points $f(a)$ et $f(b)$ and its probability is the same as the original curve $\gamma_{[ab]}$.* This is often expressed by the statement that the probability distribution is invariant under any conformal transformation:

$$P_{\mathbb{D}}(\gamma_{[ab]}) = P_{f(\mathbb{D})}\big((f \circ \gamma)_{[f(a)f(b)]}\big) \qquad (3.13)$$

which simply means that the probability transforms as a scalar under a conformal transformation. Again, the number of possible configurations of the interface being extremely large, it is not possible to construct a histogram using Monte Carlo simulations to test this relation in lattice spin models. However, one can check for example that the probability to find the interface inside a subdomain $U \subset \mathbb{D}$ is invariant under conformal transformation:

$$P_{\mathbb{D}}(\gamma_{[ab]} \subset U) = P_{f(\mathbb{D})}\big((f \circ \gamma)_{[f(a)f(b)]} \subset f(U)\big). \tag{3.14}$$

For a given lattice spin model, a first and easier step may be to check that conformal invariance holds in the sense of CFT, i.e. that correlation functions of local observables transform covariantly under conformal transformations:

$$\langle \phi_1(z_1, \bar{z}_1) \dots \phi_n(z_n, \bar{z}_n) \rangle = \left[\prod_{i=1}^{n} [f'(z_i)]^{\Delta_i} [\bar{f}'(\bar{z}_i)]^{\bar{\Delta}_i} \right] \\ \times \langle \phi_1\big(f(z_1), \bar{f}(\bar{z}_1)\big) \dots \phi_n\big(f(z_n), \bar{f}(\bar{z}_n)\big) \rangle \tag{3.15}$$

Conformal-invariance predictions for the magnetisation profile or spin-spin correlation functions for example can be checked in various geometries by applying the appropriate conformal transformation. Among the most popular conformal transformations, the logarithmic transformation

$$f(z) = \frac{L}{2\pi} \ln z \Leftrightarrow f^{-1}(z) = \exp\left(\frac{2\pi}{L} z\right) \tag{3.16}$$

maps the infinite (or half-infinite) cylinder with perimeter L onto the plane (or half-plane). Thermodynamical averages are efficiently computed in this geometry using transfer matrices. The **Cayley function**

$$f(z) = \frac{1}{i} \frac{z-1}{z+1} \Leftrightarrow f^{-1}(z) = \frac{1+iz}{1-iz} \tag{3.17}$$

maps the half-disc onto the half-plane. Monte Carlo simulations are usually performed in rectangular systems. The **Schwarz-Christoffel transformation**

$$f(z) = \operatorname{sn} K(k)z \Leftrightarrow f^{-1}(z) = F(z, k)/K(k) \tag{3.18}$$

(where sn is an elliptic function) maps the square domain $[-1; 1] \times [0; 2]$ onto the half-plane (Fig. 3.9). The constant k is given by

$$k = 4 \left[\frac{\sum_{p=0}^{+\infty} e^{-2\pi(p+1/2)^2}}{1 + 2\sum_{p=1}^{+\infty} e^{-2\pi p^2}} \right]^2 \simeq 0.1716, \tag{3.19}$$

$K(k) \simeq 1.582$ is the complete elliptic integral of first kind and $F(z, k)$ the incomplete elliptic integral of first kind. The Schwarz-Christoffel transformation can be adapted to map rectangular domains onto the upper half-plane.

In the particular case of **CFT minimal model**, a connection has been made with SLE_κ [3, 4, 46], expressed in particular by the relation

$$c = \frac{(3\kappa - 8)(6 - \kappa)}{2\kappa} \tag{3.20}$$

Fig. 3.9 Mapping of a square lattice inside the square domain $[-1; 1] \times [0; 2]$ onto the half-plane by the Schwarz-Christoffel conformal transformation (3.18). The *square corners* are mapped on the *real axis* at the points ± 1 and $\pm 1/k$

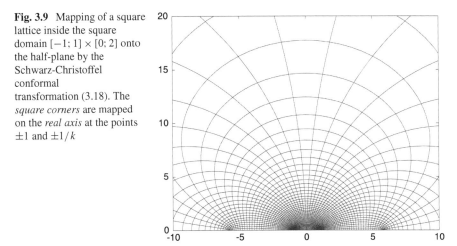

between the central charge c and the SLE parameter κ. The pure q-state Potts model with $q \leq 4$ belongs to this class of CFT. The number of states q is known to be related to the central charge by

$$q = 4\cos^2\left(\frac{\pi}{m+1}\right), \qquad c = 1 - \frac{6}{m(m+1)} \tag{3.21}$$

As a consequence, one expects to observe SLE_κ with the parameters

$$\kappa_1 = \frac{4}{1 - \frac{1}{\pi}\operatorname{Arccos}(\sqrt{q}/2)}, \qquad \kappa_2 = 16/\kappa_1 \tag{3.22}$$

These two values are associated to the hull of Fortuin-Kasteleyn clusters and their external perimeter [20]. The latter has been conjectured to behave as spin clusters. Numerical estimates of the fractal dimension of both external perimeter of FK clusters and of spin clusters are in excellent agreement [59].

Extensions to non-minimal CFTs, i.e. field theories with not only conformal symmetry but also some other ones, have been considered. Some results were obtained for example in the case of the clock model which possesses a $\mathbb{Z}_N$ additional symmetry and is related to parafermionic theories [40–42]. The Ashkin-Teller model, whose critical behaviour on the self-dual line is described by the so-called orbifold CFT, was shown to be incompatible with SLE_κ for all points of this line apart from the Ising and 4-state Potts model points [11].

3.3.3 Driving Function

Loewener's main idea was that a curve γ in the upper half-plane $\mathbb{H}$, could be studied through the (conformal) **uniformising map**

$$g_t : \mathbb{H} - \mathbb{K} \longrightarrow \mathbb{H} \tag{3.23}$$

where $\mathbb{K}$ is the hull of the curve, i.e. the domain enclosing both the curve itself and all regions inside closed loops formed when the curve touches itself. The image $g_t(\gamma)$ of the curve is on the real axis and its tip is sent to the origin: $g_t(\gamma(t)) = 0$. Under local growth of the curve, the uniformising map evolves as

$$\frac{dg_t}{dt} = \frac{2}{g_t(z) - \xi_t}, \qquad g_0(z) = z \tag{3.24}$$

where the **driving function** ξ_t is the image of the tip $\gamma_{[t,t+dt]}$ under g_t^{-1}. The factor two in the numerator comes from the hydrodynamics normalisation condition $g_t(z) \underset{z \to +\infty}{\sim} z$. One can adopt the reverse point of view: given the driving function ξ_t, one can construct a curve $\gamma(t)$ starting at O. The postulate of Markovian property and conformal invariance implies that the driving function ξ_t should be Brownian:

$$\xi_t = a + \sqrt{\kappa} B_t \tag{3.25}$$

where B_t is standard **Brownian motion** with a vanishing average and a variance equal to unity. Left-right symmetry implies $a = 0$ so that we are left with only one parameter: the **diffusion constant** κ of the Brownian motion.

In the case of lattice spin models, SLE can be tested in the following way [7]. Given a spin configuration sampled by the Monte Carlo simulation, the interface is constructed as a set of N lattice points $\{z_i\}$ with $i = 1, \ldots, N$. Since numerical calculations are not performed in the half-plane, a first conformal transformation (logarithmic map, Schwarz-Christoffel, ...) has to be applied to map the geometry used in the simulation onto the half-plane. The N points z_i are then replaced by complex coordinates in the half-plane. The idea is then to reconstruct the driving function ξ_t iteratively. One consider the first step of the interface, i.e. from O to z_1. Using the Markovian property, the uniformising map is decomposed at

$$g_t(z) = g_{\Delta t} \circ g_{t-\Delta t}(z) \tag{3.26}$$

The first map $g_{\Delta t}$ sends the first step of the interface onto the real axis, leaving thus a curve made of $N - 1$ steps in the half-plane. Since the conformal transformation that removes a vertical tip extending from $z = x_0$ to $z = x_0 + iy_0$ is (see Fig. 3.10)

$$f(z) = x_0 + \sqrt{(z - x_0)^2 + y_0^2} \tag{3.27}$$

a popular choice is to take

$$g_{\Delta t}(z) = \text{Re}(z_1) + \sqrt{(z - \text{Re}(z_1))^2 + (\text{Im} \, z_1)^2} \tag{3.28}$$

which means that we decompose the segment joining O and z_1 into first a horizontal segment $[0; \text{Re} \, z_1]$ and then a vertical one $[\text{Re} \, z_1; z_1]$. Since the driving function ξ_t is real, one reads $\xi_t = \text{Re} \, z_1$ and to recover (3.24), the time increment should be chosen as $\Delta t = (\text{Im} \, z_1)^2/4$ ($2\Delta t$ is the so-called capacity of SLE). One can check that the derivative of $g_{\Delta t}$ with respect to Δt gives indeed (3.24). The conformal transformation (3.28) must be applied to the $N - 1$ other points of the interface. This algorithm is repeated N times. More details on this algorithm can be found for example in Ref. [2]. The driving function ξ_t is averaged over a large number of spin

Fig. 3.10 Mapping of a
square lattice by the
conformal transformation
(3.27) with the values $x_0 = 1$
and $y_0 = 4$. The vertical tip
extending from $z = 1$ to
$z = 1 + 4i$ has been mapped
onto the *real axis*. The square
lattice around has been
stretched out to fill the gap

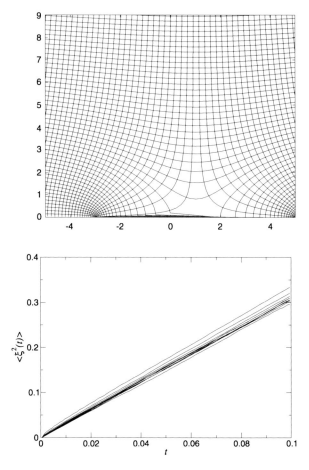

Fig. 3.11 Average square
displacement $\langle \xi_t^2 \rangle$ of the
Brownian motion extracted
from the interface between
spin clusters in the pure Ising
model. The different *curves*
correspond to different lattice
sizes: $L = 128, 160, 192,$
$224, 256, 384, 512$ and 768
(from *top* to *bottom*)

configurations sampled by the Monte Carlo simulation. If the curve is SLE_κ, one
expects the driving function ξ_t to be a Brownian motion without drift. It should be
checked that after averaging over a sufficient large number of spin configurations
(see Figs. 3.11 and 3.12)

$$\langle \xi_t \rangle = 0, \qquad \langle \xi_t^2 \rangle = \kappa t \qquad (3.29)$$

One can also check that the even moments $\langle \xi_t^{2n} \rangle$ are compatible with a Gaussian
distribution for ξ_t. To be complete, the statistical independence of the driving function ξ_t at different times t should be tested. An efficient way to do this is to compute
the correlation functions of the increments $\xi_{t+\Delta t} - \xi_t$ and check that they fall off
rapidly.

The algorithm presented above is slow. As will be discussed in the next section,
the number of points of the interface behaves as L^{d_f} with the lattice size L. For
each point of the interface, a conformal map $g_{\Delta t}$ is determined and applied to all
remaining points. The CPU time required for this operation scales therefore as L^{2d_f}.
On the other hand, a Monte Carlo iteration requires a whole sweep of the lattice and

Fig. 3.12 Effective diffusion constant κ computed by a linear fit of $\langle \xi_t^2 \rangle$ with t for the interface between spin clusters in the pure Ising and the 3-state Potts models. The *dashed lines* are the expected theoretical values $\kappa = 3$ and $10/3$

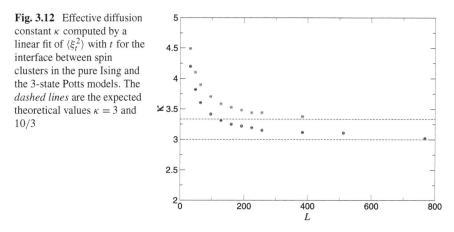

demands thus a CPU time scaling as L^2. The spin configurations sampled by the simulation being correlated, the number of Monte Carlo steps required to obtain two statistically independent configurations is of the order of $\tau \sim L^z$ where z is the **dynamical exponent** at the critical point. As a consequence, the CPU time of the Monte Carlo part of the code scales as L^{2+z}. Using cluster algorithms, z is close to zero and smaller than $2d_f - 2$. The computation of the driving function becomes rapidly the more time-consuming part of the code as the lattice size is increased. To circumvent this problem, Kennedy has proposed an improved algorithm [31–33]. Instead of applying the conformal map $g_{\Delta t}(z)$ (3.28) to all other points z_j with $j > i$ when the point z_i is removed, the conformal map is approximated by the Laurent series

$$g_{\Delta t}(1/z) = 1/z + \sum_{k=0}^{M} c_k z^k \qquad (3.30)$$

where M is a cut-off that needs to be tuned. The coefficients c_k can be stored and manipulated by the computer. At each iteration the full uniformising map

$$g_t = g_{\Delta t_n}(1/z) \circ g_{\Delta t_{n-1}} \circ \ldots \circ g_{\Delta t_1}(1/z) \qquad (3.31)$$

is updated in the sense that (3.28) is composed with (3.30) and the result is put under the same form as (3.30). The conformal transformation g_t is applied only to the first point to be removed.

3.3.4 Finite-Size Scaling of Length and Winding Number

The SLE trace being fractal, its length ℓ is expected to scale in a domain of size L as

$$\ell \sim L^{d_f} \qquad (3.32)$$

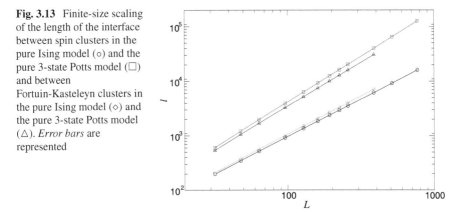

Fig. 3.13 Finite-size scaling of the length of the interface between spin clusters in the pure Ising model (o) and the pure 3-state Potts model (□) and between Fortuin-Kasteleyn clusters in the pure Ising model (◇) and the pure 3-state Potts model (△). *Error bars* are represented

The **fractal dimension** d_f is related to the diffusion constant κ by [6, 45]

$$d_f = 1 + \frac{\kappa}{8} \tag{3.33}$$

for $\kappa \leq 8$ while $d_f = 2$ for $\kappa > 8$. In Figs. 3.13 and 3.14, an example of numerical determination of κ using Eqs. (3.32) and (3.33) is presented. Values in the literature are reported in Table 3.1. One should remember that d_f is just a number. Measuring d_f does not prove that SLE holds! However, if the central charge c is known, the validity of Eq. (3.20) may be tested. Moreover, one can check that the fractal dimension d'_f of the external perimeter is $1 + \frac{\kappa'}{8}$ with $\kappa' = 16/\kappa$ or equivalently $(d_f - 1)(d'_f - 1) = 4$. These tests have been performed for the Fortuin-Kasteleyn random clusters by means of Monte Carlo simulations for integer values of the number of states [1] as well as for non-integer ones [25, 59]. Because they can cross and branch, an infinite spectrum of fractal dimensions may be defined for the spin clusters [18, 19].

The diffusion constant κ can also be extracted from the statistics of the winding number of the curve around long cylinders. When the cylinder is conformally mapped onto the half-plane, the winding number is related to the polar angle. The latter can be estimated as the sum of the angles between each pair of successive edges that form the interface [57]. The variance of the winding number θ is expected to behave as $\langle \theta^2 \rangle \sim \kappa \ln L$. Example for the 2D Ising model can be found in [47].

3.3.5 Left-Passage Probability

Schramm has shown that the probability that the curve $\gamma(t)$ passes at the left of a given point $z = x + iy$ of the upper half-plane is given by the formula [49]

$$P_{\text{Left}}(z) = \frac{1}{2} + \frac{\Gamma(4/\kappa)}{\sqrt{\pi}\,\Gamma((8-\kappa)/2\kappa)} \frac{x}{y} \, {}_2F_1\left(\frac{1}{2}, \frac{4}{\kappa}, \frac{3}{2}, -\frac{x^2}{y^2}\right) \tag{3.34}$$

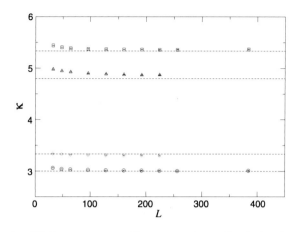

Fig. 3.14 Effective diffusion constants κ with respect to the smallest lattice size used during the fit of the length of the interface with the lattice size L. The *symbols* correspond to the interface between spin clusters in the pure Ising model (○) and the pure 3-state Potts model (□) and between Fortuin-Kasteleyn clusters in the pure Ising model (◇) and the pure 3-state Potts model (△). The *dashed lines* correspond to the expected values, resp. 3, 10/3, 16/3 and 24/5

Table 3.1 Fractal dimensions d_f and corresponding SLE diffusion constants κ of the interfaces between spin clusters in the pure Ising and 3-state Potts models, according to [24]

Model	d_f	κ	κ (Theory)
Ising	1.372(2)	2.976(20)	3
3-state Potts	1.399(2)	3.192(20)	10/3

where $_2F_1$ is the hypergeometric function. It can be seen that the probability depends only on $x/y = \cot\theta$, i.e. on the polar angle θ as demanded by dilatation-invariance (Fig. 3.15).

Schramm's formula (3.34) is particularly appropriate for a test by Monte Carlo simulation. At each Monte Carlo simulation, the interface is first identified. The left of the interface is then filled and for each site visited during the filling, the histogram of the number of visit is updated. Instead of trying to interpolate the data $P(x, y)$ with Schramm's formula $P_{\text{Left}}(x, y)$, the square deviation

$$\chi^2(x, y) = \left[P(x, y) - P_{\text{Left}}(x, y) \right]^2, \qquad \overline{\chi^2} = \frac{1}{L^2} \sum_{x,y} \chi^2(x, y) \qquad (3.35)$$

of the data with Schramm's formula is computed for various values of κ and its minimum is searched (see Fig. 3.16). Compatibility of the data with the Schramm formula means that $\sqrt{\chi^2(x, y)}$ is smaller than the error bar on the left-passage probability. Lattice effects manifest themselves close to the boundaries and especially at the starting and ending point of the interface. The Schramm formula usually works poorly there. However, Schramm's formula allows for very accurate, and less sensible to finite-size effects, estimates of κ because for each lattice size L, statistical

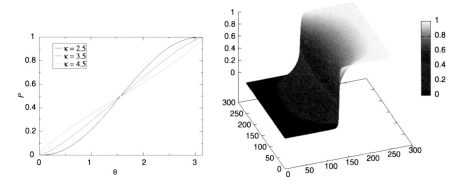

Fig. 3.15 On the *left*, exact left-passage probabilities in the half-plane as a function of the polar angle θ. The different *curves* correspond to different values of the diffusion constant κ. On the *right*, average left-passage probability calculated numerically for the pure Ising model on a 256×256 square lattice

Fig. 3.16 Mean-square deviation $\overline{\chi^2}$ from the Schramm formula for the pure Ising model with respect to κ. The different *curves* correspond to lattices sizes $L = 64, 96, 128, 160, 192, 224, 256, 384, 512,$ and 768 (from *top* to *bottom*). For comparison, the sum of the quadratic error of the left-passage probability on all sites are represented as *error bars*

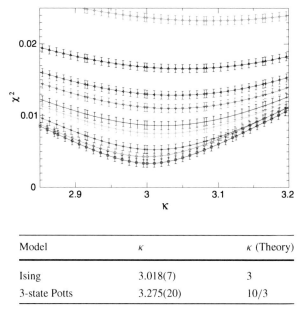

Table 3.2 SLE diffusion constants κ obtained from the Schramm formula in the pure Ising and 3-state Potts models, according to [24]

Model	κ	κ (Theory)
Ising	3.018(7)	3
3-state Potts	3.275(20)	10/3

fluctuations are reduced when taking into account all points of the lattice. Example of values of κ in literature obtained from the Schramm formula is given in Table 3.2.

3.4 SLE in Random Lattice Spin Models

As previously discussed, the question whether interfaces in random systems are SLE traces or not, is still largely open. In particular, Domain Markov property does not hold trivially as for pure models and it is unclear why it should hold for random systems. Moreover, conformal invariance is broken for a particular disorder realisation but may be restored after average over randomness. However, recent numerical simulations performed in the last five years provide evidences that interfaces in some random systems, frustrated or not, may be SLE traces.

3.4.1 Random Potts Model

The **random-bond Potts model** is defined by the hamiltonian

$$\mathcal{H} = -\sum_{(i,j)} J_{ij}\delta_{\sigma_i,\sigma_j}, \quad \sigma_i \in \{0,\ldots,q-1\} \tag{3.36}$$

where the exchange couplings J_{ij} are quenched positive random variables. On the square lattice with the binary distribution

$$P(J_{ij}) = \frac{1}{2}\left[\delta(J_{ij}-J_1) + \delta(J_{ij}-J_2)\right] \tag{3.37}$$

self-duality arguments allow for the determination of the critical line:

$$\left(e^{J_1}-1\right)\left(e^{J_2}-1\right) = q, \quad r = J_1/J_2 \tag{3.38}$$

For $q \leq 4$, the pure model ($r=1$) displays a second-order phase transition with q-dependent critical exponents. As already mentioned, to each value of the number of states q correspond a CFT with a central charge going from $c=1/2$ for $q=2$ to $c=1$ for $q=4$. According to the **Harris criterion** [26], *disorder coupled to the energy density is marginally irrelevant for $q=2$ and relevant for $q>2$*. An expansion in powers of $q-2$ of critical exponents for the random q-state Potts model was determined by Renormalisation-group methods applied to the Coulomb gas representation of the Potts model [17, 36, 37]. The values were later confirmed by transfer matrix calculations [10, 29]. In the regime $q>4$, the pure Potts model displays a first-order phase transition which is smoothed by randomness and turned into a continuous transition with non-trivial, q-dependent, critical exponents [13].

Numerical simulations provided evidences that conformal invariance is restored in the continuum limit after suitable disorder averaging. As shown by transfer-matrix calculations [10, 29], spin-spin correlation functions in the strip geometry decay exponentially as expected in order to recover an algebraic decay in the infinite plan after the conformal map (3.16). The central charge was extracted from the finite-size scaling of the free-energy density on the strip. If conformal symmetry holds, the CFT behind the random Potts model cannot be a minimal one. Besides conformal symmetry, the model is indeed expected to be invariant under permutation of replicas [17, 36, 37] and the CFT is not unitary. As a consequence, the relation (3.20) is not expected to hold in the case of the random Potts model. From a purely numerical point of view, the central charge turned out to be very useful to tune the disorder ratio $r = J_1/J_2$ in order to minimise the cross-over effects with the pure and infinite-disorder fixed points. On square lattices, the conformal predictions

Table 3.3 Fractal dimensions d_f and corresponding SLE diffusion constants κ of the interfaces between Fortuin-Kasteleyn and spin clusters in the 3-state random Potts model, as determined from renormalisation-group (RG) methods, transfer matrix calculations and Monte Carlo simulations, according to [30]

	Fortuin-Kasteleyn		Spin clusters	
	d_f	κ	d_f	κ
RG	1.61433	4.9146		
Monte Carlo	1.614(3)	4.912(24)	1.401(3)	3.208(24)
Transfer Matrix	1.615(2)	4.920(16)		

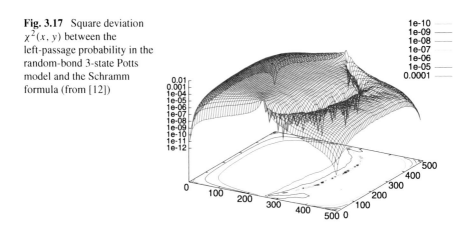

Fig. 3.17 Square deviation $\chi^2(x, y)$ between the left-passage probability in the random-bond 3-state Potts model and the Schramm formula (from [12])

for the magnetisation profile and spin-spin correlations functions have been tested using Monte Carlo simulations [14, 15].

Renormalisation-group calculations were recently extended to the study of the geometry of the critical curves in the regime $q \leq 4$ [30]. The fractal dimension of the interfaces between Fortuin-Kasteleyn clusters was predicted to be 1.61433 for the random 3-state Potts model. This value was confirmed by finite-size scaling of the average length of the interface computed by means of transfer-matrix calculations and Monte Carlo simulations, see Table 3.3. Additionally, the fractal dimension of the interface between spin clusters was estimated numerically. The two corresponding values of κ are compatible with the relation $\kappa\kappa' = 16$.

The Schramm formula for the left-passage probability was latter tested in a square geometry [12] (see Figs. 3.17 and 3.18). A systematic deviation was observed and interpreted as due to the fluctuating Boundary Conditions. Although smaller, this deviation exists for the pure Potts model too. However, by restricting the calculation of the deviation to the region in the neighbourhood of the fixed part of the boundary, *the diffusion constant κ takes a value compatible with the one found from the fractal dimension, both in the pure and random Potts model*. In the latter, the diffusion constant is estimated to be $\kappa = 3.245(10)$.

Fig. 3.18 *Error bars* on the
left-passage probability in the
random-bond 3-state Potts
model (from [12])

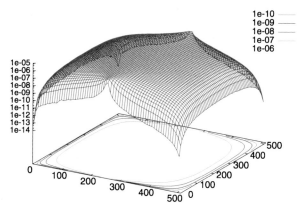

3.4.2 SOS Model on a Random Substrate

The two-dimensional **Solid-On-Solid model** (SOS model) on a disordered substrate [21, 50] is defined by the Gaussian hamiltonian

$$H = \sum_{(i,j)} (h_i - h_j)^2 \tag{3.39}$$

where $h_i = d_i + n_i$ is the total height at site i. The height of the substrate d_i is a
quenched random variable uniformly distributed in $[0; 1]$. The degrees of freedom of
the system n_i can take any integer values (positive or negative). This model displays
a phase transition between a rough high-temperature phase (with roughness $w^2 \sim
\ln L$) and a super-rough glassy phase (with roughness $w^2 \sim (\ln L)^2$). The system
is initially prepared with fixed boundary conditions $n = 0$. The ground state n_i^0 is
then numerically determined using a minimal cost flow algorithm. On one half of
the boundary, the height are then changed to $n_i = 1$. The new ground state displays
an interface between a region where the ground state has not changed and a region
where it has been shifted, i.e. $n_i = n_i^0 + 1$, due to the new boundary conditions.

Using the finite-size scaling of the average length of the interface, the fractal dimension was estimated to be $d_f \simeq 1.25(1)$ which implies $\kappa \simeq 2.00(8)$ according to
(3.33). However, the smallest deviation of left-passage probability to the Schramm
formula (3.34) is observed for $\kappa \simeq 4.00(1)$, for both a square domain and a half-
circle. The incompatibility of the two estimates led the authors to the conclusion
that *the random SOS interface cannot be described by SLE*.

3.4.3 Ising Spin Glasses

The **Edwards-Anderson model** of **spin glasses** is defined by the hamiltonian [22]

$$\mathcal{H} = -\sum_{(i,j)} J_{ij}\sigma_i\sigma_j \tag{3.40}$$

where the coupling constants J_{ij} are quenched random variables that can take both
positive and negative values. The latter produce frustration in the system leading to

the very slow dynamics characteristic of glasses. Popular choices for the probability distribution of these couplings are the bimodal distribution $\pm J$ and the Gaussian distribution. In dimension $d = 2$, the correlation length ξ is known to diverge at zero temperature.

Although the inhomogeneous configuration of coupling constants J_{ij} breaks conformal invariance, it seems to be restored as long as one consider average quantities. A large-scale Monte Carlo simulation of the two-dimensional Ising spin glass (ISG) provided evidences in that sense [2]. The shape of the lattice was chosen to be a strip whose boundary conditions were chosen to be periodic in one direction, say horizontal, and open in the other one, i.e. the vertical one. Two cuts are then made vertically from the upper and lower boundaries (i.e. bonds are removed or equivalently J_{ij} are set to zero) leaving a slit of untouched bonds around the centre of that column (see Fig. 1 of [2]). Since these cuts can be removed by a conformal transformation, the probability that the domain wall crosses n times the slit is the same that the probability of crossing of any other column without cuts. To induce the domain wall, a ground state spin configuration $\sigma_i^{(0)}$ is first determined numerically and then all coupling constants are flipped in one column, i.e. $J_{ij} \rightarrow -J_{ij}$. Since there are two ground states related by a global spin flip, in one part of the lattice the spins remains unchanged after the operation, i.e. $\sigma_i = \sigma_i^{(0)}$, while in the other part, $\sigma_i = -\sigma_i^{(0)}$. The domain wall is defined as the curve on the dual lattice separating the region where $\sigma_i = \sigma_i^{(0)}$ from the one where $\sigma_i = -\sigma_i^{(0)}$, i.e. as the curve crossing the links between sites i and j for which $\sigma_i \sigma_i^{(0)} = -\sigma_j \sigma_j^{(0)}$. In contradistinction to previously discussed boundary conditions, the ending points of the domain wall is not fixed in this case. For this reason, the domain wall was said to be floating [8]. By means of Monte Carlo simulations, the probabilities that the domain wall crosses n times resp. the slit $(p_1'(n)$ and another column $p_2'(n))$ were measured. The probabilities of no-crossing, i.e. $p_1'(0)$ and $p_2'(0)$, converge very slowly only to the same value in the infinite-lattice limit but the authors were able to conclude that the probabilities are the same for an even number n of crossings. To test specifically SLE, the authors extracted the driving function using the algorithm presented in Sect. 3.3.3. For sufficiently small values of t, the variance $\langle \xi_t^2 \rangle$ is linear (see Fig. 3.19) and allows for the estimation of $\kappa \simeq 2.1$, consistent with a previous estimate of $d_f \simeq 1.27$ [27] for a Gaussian distribution of couplings. By identifying the conformal weight that could be compatible with this value, the authors suggested a relation between κ and the stiffness exponent θ (describing the finite-size power-law behaviour of the energy of the domain wall):

$$d_f = 1 + \frac{3}{4(3+\theta)} \Leftrightarrow \kappa = \frac{6}{3+\theta} \tag{3.41}$$

Note that the first equality (but not the second) holds for the SOS model on a random substrate (even though the interface does not seem to be described by SLE for this model) since the stiffness exponent is $\theta = 0$ ($\Delta E \sim \ln L$) and the fractal dimension was estimated to be $d_f \simeq 1.25$.

Further numerical tests of SLE in ISG were presented a few months latter [8]. The fractal dimension $d_f = 1.28(1)$ was confirmed. More recent (and more precise)

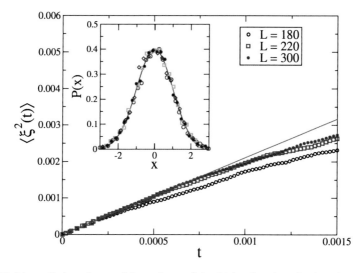

Fig. 3.19 Monte Carlo estimate of the variance of the driving function of a domain wall in the two-dimensional Ising spin glass (from [2])

estimates are compatible with this value: 1.274(2) [38], 1.275 [39] and confirm Eq. (3.41). Conformal predictions for the winding number of the floating domain wall around long cylinders are reproduced by the data with a value of the fractal dimension compatible to the above estimate. Measurements of the left-passage probability were shown to be compatible with the Schramm formula with $\kappa = 2.32(8)$ (see Fig. 3.20). The boundary conditions were periodic in one direction (in the sense that the spins on both sides of the boundary are chosen such that $\sigma_i \sigma_j$ has the same sign as J_{ij}) and anti-periodic in the other. Like in [2], the ending points of the domain wall are therefore not fixed but allowed to "float". When the starting point of the interface is fixed by changing a bond on the lower boundary, a slightly larger value 2.85(10) is obtained from the Schramm formula.

The driving function ξ_t was extracted from Monte Carlo data and analysed. Because of the strip geometry and the boundary conditions, the domain wall is not related to chordal SLE (discussed above) but to dipolar SLE [5]. In the latter, the SLE trace is grown from a fixed point of the lower boundary and stops when hitting the upper boundary for the first time. For this reason, the formulæ employed in [8] to reconstruct the driving function differ from those presented above for chordal SLE. The data is consistent with a Gaussian distribution of ξ_t and the analysis of the even moments provided compatible estimates of the diffusion constant: $\kappa = 2.24(8)$ (floating domain wall) and 2.85(10) (fixed starting point). Correlations between increments $\xi_{t+1} - \xi_t$ of the driving function were shown to decay quickly as expected for an uncorrelated Brownian noise. All these tests provided *strong evidences that conformal invariance holds for ISG and that domain walls can be described by SLE traces with $\kappa \simeq 2.3$*. However, the domain Markovian property seems not to be satisfied in small systems while the test is inconclusive for larger ones.

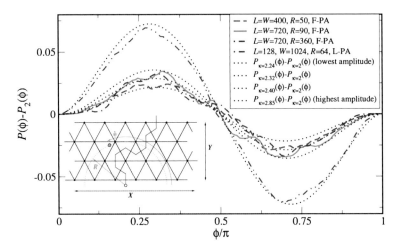

Fig. 3.20 Comparison of the left-passage probability with the Schramm formula in Ising spin glasses. The data are plotted with respect to the polar angle (from [8])

The results presented above were obtained using a Gaussian distribution of the coupling constants J_{ij}. Numerical simulations in the bimodal case are more difficult because of the high degeneracy of the ground state. Using an exact optimisation algorithm for small lattice sizes and a variant of Parallel Tempering Monte Carlo simulation for larger ones, Risau et al. estimated the fractal dimension of the domain wall to be 1.279(4) for a Gaussian distribution of couplings and 1.323(3) for a Bimodal one [44]. The left-passage probability was measured and the comparison with the Schramm formula gave an estimate $\kappa \simeq 2.23$ in the Gaussian case that is compatible with the relation (3.33) and the estimate $d_f \simeq 1.279(4)$. However, the data for the Bimodal case turned out to be fully incompatible with the Schramm formula with $\kappa \simeq 2.584$ given by (3.33) and the estimate $d_f \simeq 1.323(4)$. The relation (3.41) does not hold (see Figs. 3.21, 3.22). This discrepancy remains unexplained. As noted by the authors, the two distributions lead to different reaction of the ISG to the introduction of anti-periodic boundary conditions (aPBC). With a Gaussian distribution, a single domain wall is induced, separating one part of the system where spins have flipped and the part where they remained unchanged. In contradistinction, additional spin clusters are flipped without any energy cost in the Bimodal case. They form closed loops that may be glued onto the domain wall. If no specific procedure is applied to remove them, the numerical determination of the length of the domain wall may be erroneous.

3.4.4 Random-Field Ising Model

The **Random-field Ising model** (RFIM) is defined by the hamiltonian

$$\mathscr{H} = -\sum_{(i,j)} \sigma_i \sigma_j - \sum_i h_i \sigma_i \tag{3.42}$$

Fig. 3.21 Finite-size scaling
of the length of the domain in
the two-dimensional and
three-dimensional Ising Spin
Glass, for a Gaussian
distribution of coupling
constants (EAG) and a
bi-modal one (EAB)
(from [44])

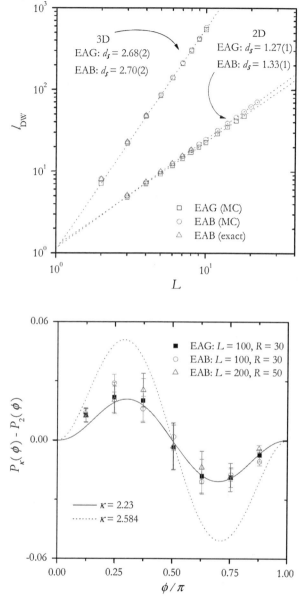

Fig. 3.22 Comparison of the
difference of the left-passage
probability with the Schramm
formula with respect to the
polar angle, for the $2D$ and
$3D$ Ising spin glass with a
Gaussian (EAG) or bi-modal
(EAB) distribution of the
coupling constants
(from [44])

where the local magnetic fields h_i are quenched random variables, usually chosen to
be Gaussian with standard deviation denoted δ. In dimension $d = 2$, the ferromagnetic order is destroyed at any temperature (the lower critical dimension is $d_c = 2$).
The system remains paramagnetic and undergoes no phase transition. However, the
size of spin clusters diverges at zero temperature for a critical value $h = h_c(\Delta)$ of
the random magnetic field.

SLE has been tested recently for this model [55]. Ground states have been numerically sampled using a fast minimum-cut/maximum flow algorithm. Interfaces were induced by fixing spins at part of the boundaries. The Domain Markov Property has been checked by comparing the left-passage probability obtained first in a domain $\mathbb{D}$ with an average restricted to the spin configurations yielding an interface along $\gamma_{[ac]}$ and second in a domain $\mathbb{D} \setminus \gamma_{[ac]}$ with a cut along $\gamma_{[ac]}$. The two probabilities should be equal if Domain Markov Property holds. Numerical calculations show a significant difference in the neighbourhood of the tip of $\gamma_{[ac]}$. However, a close inspection of the finite-size behaviour of this difference indicates a vanishing in the continuum limit. The left-passage probability in the unit disk was shown to be compatible with the Schramm formula with $\kappa \simeq 6.00(5)$ (like percolation). Crossing probabilities of percolating clusters, finite-size scaling of the interface length and variance of the driving function ξ_t corroborate this value. A bimodal distribution of the random fields has also been studied. In contradistinction to spin glasses, *the RFIM-data are in agreement with SLE predictions with $\kappa = 6$.*

References

1. Adams, D.A., Sander, L.M., Ziff, R.M.: Fractal dimensions of the Q-state Potts model for complete and external hulls. J. Stat. Mech. P03004 (2010)
2. Amoruso, C., Hartmann, A.K., Hastings, M.B., Moore, M.A.: Conformal invariance and stochastic Loewner evolution processes in two-dimensional Ising spin glasses. Phys. Rev. Lett. **97**, 267202 (2006)
3. Bauer, M., Bernard, D.: Conformal field theories of stochastic Loewner evolutions. Commun. Math. Phys. **239**, 493 (2003)
4. Bauer, M., Bernard, D.: Conformal transformations and the SLE partition function martingale. Ann. Henri Poincaré **5**, 289 (2004)
5. Bauer, M., Bernard, D., Houdayer, J.: Dipolar SLEs. J. Stat. Mech. P03001 (2005)
6. Beffara, V.: Hausdorff dimensions for SLE_6. Ann. Probab. **32**, 2606 (2002)
7. Bernard, D., Boffetta, G., Celani, A., Falkovich, G.: Inverse turbulent cascades and conformally invariant curves. Phys. Rev. Lett. **98**, 024501 (2007)
8. Bernard, D., Doussal, P.L., Middleton, A.A.: Possible description of domain walls in two-dimensional spin glasses by stochastic Loewner evolutions. Phys. Rev. B **76**, 020403 (2007)
9. Camia, F., Newman, C.M.: Critical percolation exploration path and SLE_6: a proof of convergence. Probab. Theory Relat. Fields **139**, 473 (2007)
10. Cardy, J.L., Jacobsen, J.L.: Critical behaviour of random-bond Potts models. Phys. Rev. Lett. **79**, 4063 (1997)
11. Caselle, M., Lottini, S., Rajabpour, M.A.: Critical domain walls in the Ashkin-Teller model. J. Stat. Mech. P02039 (2011)
12. Chatelain, C.: Numerical study of Schramm-Loewner evolution in the random 3-state Potts model. J. Stat. Mech. P08004 (2010)
13. Chatelain, C., Berche, B.: Finite-size scaling study of the surface and sulk critical behaviour in the random-bond eight-states Potts model. Phys. Rev. Lett. **80**, 1670 (1998)
14. Chatelain, C., Berche, B.: Tests of conformal invariance in randomness-induced second-order phase transitions. Phys. Rev. E **58**, 6899 (1998)
15. Chatelain, C., Berche, B.: Magnetic critical behavior of two-dimensional random-bond Potts ferromagnets in confined geometries. Phys. Rev. E **60**, 3853 (1999)
16. Coniglio, A.: Fractal structure of Ising and Potts clusters: exact results. Phys. Rev. Lett. **62**, 3054 (1989)

17. Dotsenko, V.S., Picco, M., Pujol, P.: Renormalisation-group calculation of correlation functions for the $2D$ random bond Ising and Potts models. Nucl. Phys. B **455**, 701 (1995)
18. Dubail, J., Jacobsen, J.L., Saleur, H.: Bulk and boundary critical behaviour of thin and thick domain walls in the two-dimensional Potts model. J. Stat. Mech. P12026 (2010)
19. Dubail, J., Jacobsen, J.L., Saleur, H.: Critical exponents of domain walls in the two-dimensional Potts model. J. Phys. A, Math. Theor. **43**, 482002 (2010)
20. Duplantier, B.: Conformally invariant fractals and potential theory. Phys. Rev. Lett. **84**, 1363 (2000)
21. Duxbury, P.M.: Exact computations test stochastic Loewner evolution and scaling in glassy systems. J. Stat. Mech. N09001 (2009)
22. Edwards, S.F., Anderson, P.W.: Theory of spin glasses. J. Phys. F, Met. Phys. **5**, 965 (1975)
23. Fortuin, C.M., Kasteleyn, P.W.: On the random-cluster model. I. Introduction and relation to other models. Physica **57**, 536 (1972)
24. Gamsa, A., Cardy, J.L.: Schramm-Loewner evolution in the three-state Potts model: a numerical study. J. Stat. Mech. P08020 (2007)
25. Gliozzi, F., Rajabpour, M.A.: Conformal curves in the Potts model: numerical calculation. J. Stat. Mech. L05004 (2010)
26. Harris, A.B.: Effect of random defects on the critical behaviour of Ising models. J. Phys. C, Solid State Phys. **7**, 1671 (1974)
27. Hartmann, A.K., Young, A.P.: Large-scale low-energy excitations in the two-dimensional Ising spin glass. Phys. Rev. B **66**, 094419 (2002)
28. Ising, E.: Beitrag zur Theorie des Ferromagnetismus. Z. Phys. **31**, 253 (1925)
29. Jacobsen, J.L., Cardy, J.L.: Critical behaviour of random-bond Potts models: a transfer matrix study. Nucl. Phys. B **515**, 701 (1998)
30. Jacobsen, J.L., Doussal, P.L., Picco, M., Santachiara, R., Wiese, K.J.: Critical interfaces in the random-bond Potts model. Phys. Rev. Lett. **102**, 070601 (2009)
31. Kennedy, T.: A fast algorithm for simulating the chordal Schramm-Loewner evolution. J. Stat. Phys. **128**, 1125 (2007)
32. Kennedy, T.: Computing the Loewner driving process of random curves in the half plane. J. Stat. Phys. **131**, 803 (2008)
33. Kennedy, T.: Numerical computations for the Schramm-Loewner evolution. J. Stat. Phys. **137**, 839 (2009)
34. Lawler, G.F., Schramm, O., Werner, W.: Conformal invariance of planar loop-erased random walks and uniform spanning tress. Ann. Probab. **32**, 939 (2004)
35. Lawler, G.F., Schramm, O., Werner, W.: On the scaling limit of planar self-avoiding walk. In: Fractal Geometry and Applications: A Jubilee of Benoît Mandelbrot, Part 2, p. 339. Am. Math. Soc., Providence (2004)
36. Ludwig, A.W.W.: Critical behaviour of the two-dimensional random q-state Potts model by expansion in $(q - 2)$. Nucl. Phys. B **285**, 97 (1987)
37. Ludwig, A.W.W., Cardy, J.L.: Perturbative evaluation of the conformal anomaly at new critical points with applications to random systems. Nucl. Phys. B **285**, 687 (1987)
38. Melchert, O., Hartmann, A.K.: Fractal dimension of domain walls in two-dimensional Ising spin glasses. Phys. Rev. B **76**, 174411 (2007)
39. Melchert, O., Hartmann, A.K.: Scaling behavior of domain walls at the $T = 0$ ferromagnet to spin-glass transition. Phys. Rev. B **79**, 184402 (2009)
40. Picco, M., Santachiara, R.: Numerical study on Schramm-Loewner evolution in nonminimal conformal field theories. Phys. Rev. Lett. **100**, 015704 (2008)
41. Picco, M., Santachiara, R.: Critical interfaces of the Ashkin-Teller model at the parafermionic point. J. Stat. Mech. P07027 (2010)
42. Picco, M., Santachiara, R., Sicilia, A.: Geometrical properties of parafermionic spin models. J. Stat. Mech. P04013 (2009)
43. Potts, R.B.: Some generalized order-disorder transformations. Math. Proc. Camb. Philos. Soc. **48**, 106 (1952)
44. Risau-Gusman, S., Romá, F.: Fractal dimension of domain walls in the Edwards-Anderson spin glass model. Phys. Rev. B **77**, 134435 (2008)

45. Rohde, S., Schramm, O.: Basic properties of SLE. Ann. Math. **161**, 879 (2005)
46. Rushkin, I., Bettelheim, E., Gruzberg, I.A., Wiegmann, P.: Critical curves in conformally invariant statistical systems. J. Phys. A, Math. Theor. **40**, 2165 (2007)
47. Saberi, A.A.: Thermal behaviour of spin clusters and interfaces in the two-dimensional Ising model on a square lattice. J. Stat. Mech. P07030 (2009)
48. Schramm, O.: Scaling limits of loop-erased random walks and uniform spanning trees. Isr. J. Math. **118**, 221 (2000)
49. Schramm, O.: A percolation formula. Electron. Commun. Probab. **6**, 115 (2001)
50. Schwarz, K., Karrenbauer, A., Schehr, G., Rieger, H.: Domain walls and chaos in the disordered SOS model. J. Stat. Mech. P08022 (2009)
51. Smirnov, S.: Critical percolation in the plane: conformal invariance, Cardy's formula, scaling limits. C. R. Acad. Sci. Paris **333**, 239 (2001)
52. Smirnov, S.: Conformal invariance in random cluster models. I. Holomorphic fermions in the Ising model. Ann. Math. **172**, 1435 (2010)
53. Stanley, H.E.: Cluster shapes at the percolation threshold: and effective cluster dimensionality and its connection with critical-point exponents. J. Phys. A, Math. Gen. **10**, 211 (1977)
54. Stevenson, J.D., Weigel, M.: Domain walls and Schramm-Loewner evolution in the random-field Ising model. Europhys. Lett. **95**, 40001 (2011)
55. Stevenson, J.D., Weigel, M.: Percolation and Schramm-Loewner evolution in the $2D$ random-field Ising model. Comput. Phys. Commun. **182**, 1879 (2011)
56. Swendsen, R.H., Wang, J.S.: Nonuniversal critical dynamics in Monte Carlo simulations. Phys. Rev. Lett. **58**, 86 (1987)
57. Wieland, B., Wilson, D.B.: Winding angle variance of Fortuin-Kasteleyn contours. Phys. Rev. E **68**, 056101 (2003)
58. Wolff, U.: Collective Monte Carlo updating for spin systems. Phys. Rev. Lett. **62**, 361 (1989)
59. Zatelepin, A., Shchur, L.: Duality of critical interfaces in Potts model: numerical check. arXiv:1008.3573 (2010)

Chapter 4
Loop Models and Boundary CFT

Jesper Lykke Jacobsen

4.1 Models and Transformations

The guiding principle of these lectures is to use a few simple models as exploratory tools for presenting a whole range of exact techniques within two-dimensional CFT.

For that reason, we focus on two models with a particularly rich physical and mathematical content: the *Q-state Potts model* and the O(n) *model*. In this first section we shall see that these models can be formulated in several equivalent ways, in terms of different degrees of freedom. Some of these degrees of freedom are defined locally (spin, arrows, heights), and some are spatially extended objects (clusters, loops). The extended degrees of freedom make manifest the geometrical content of the underlying models, furnishing in the same time a range of concrete physical applications and an intuitive picture of the long-range correlation that we aim at describing.

From a physical perspective, these different formulations are linked by a series of ingenuous transformations. On the mathematical side, the models admit an algebraic formulation, and the various formulations are mirrored by the existence of different representation of the same algebra. The algebra underlying the Potts model is the celebrated *Temperley-Lieb* (TL) *algebra*, and the O(n) model is described by a close cousin, which is the *dilute* TL *algebra*.

4.1.1 Potts Model

Let $G = (V, E)$ be an arbitrary **graph** with **vertex set** V and **edge set** E. The Q-state **Potts model** is initially defined by assigning a variable σ_i, henceforth called

J.L. Jacobsen (✉)
Laboratoire de Physique Théorique de l'École Normale Supérieure, 24 rue Lhomond, 75005 Paris, France
e-mail: jesper.jacobsen@ens.fr

M. Henkel, D. Karevski (eds.), *Conformal Invariance: an Introduction to Loops, Interfaces and Stochastic Loewner Evolution*, Lecture Notes in Physics 853, DOI 10.1007/978-3-642-27934-8_4, © Springer-Verlag Berlin Heidelberg 2012

a **spin**, to each vertex $i \in V$. Each spin can take Q different values, by convention chosen as $\sigma_i = 1, 2, \ldots, Q$. We denote by σ the collection of all spin variables on the graph. Two spins i and j are called **nearest neighbours** if they are incident on a common edge $e = (ij) \in E$. In any given configuration σ, a pair of nearest-neighbour spins is assigned an energy $-J$ if they take identical values, $\sigma_i = \sigma_j$. The Hamiltonian (dimensionless energy functional) of the Potts model is thus

$$\mathcal{H} = -K \sum_{(i,j) \in E} \delta(\sigma_i, \sigma_j), \tag{4.1}$$

where the Kronecker delta function is defined as

$$\delta(\sigma_i, \sigma_j) = \begin{cases} 1 & \text{if } \sigma_i = \sigma_j \\ 0 & \text{otherwise} \end{cases} \tag{4.2}$$

and $K = J/k_B T$ is a dimensionless coupling constant (interaction energy).

The thermodynamical information about the Potts model is encoded in the **partition function**

$$Z = \sum_{\sigma} e^{-\mathcal{H}} = \sum_{\sigma} \prod_{(ij) \in E} e^{K\delta(\sigma_i, \sigma_j)} \tag{4.3}$$

and in various correlation functions. By a **correlation function** we understand the probability that a given set of vertices are assigned fixed values of the spins.

In the ferromagnetic case $K > 0$ the spins tend to align at low temperatures ($K \gg 1$), defining a phase of ferromagnetic order. Conversely, at high temperatures ($K \ll 1$) the spins are almost independent, leading to a paramagnetic phase where entropic effects prevail. On physical grounds, one expects the two phases to be separated by a critical point K_c where the effective interactions between spins become long-ranged.

For certain regular planar lattices K_c can be determined exactly by duality considerations. Moreover, K_c will turn out to be the locus of a second-order phase transition if $0 \leq Q \leq 4$. In that case, the Potts model enjoys conformal invariance in the limit of an infinite lattice, allowing its critical properties to be determined exactly by a variety of techniques. These properties turn out to be *universal*, i.e., independent of the lattice used for defining the model microscopically.

4.1.1.1 Fortuin-Kasteleyn Cluster Representation

The initial definition (4.1) of the Potts model requires the number of spins Q to be a positive integer. It is possible to rewrite the partition function and correlation functions so that Q appears only as a parameter. This makes its possible to assign to Q arbitrary real (or even complex) values.

Notice first that by (4.2) we have the identity

$$e^{K\delta(\sigma_i, \sigma_j)} = 1 + v\delta(\sigma_i, \sigma_j), \tag{4.4}$$

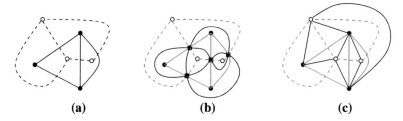

Fig. 4.1 (**a**) A planar graph G (*black circles* and *solid lines*) and its dual graph G^* (*white circles* and *dashed lines*). (**b**) The medial graph $\mathcal{M}(G) = \mathcal{M}(G^*)$. (**c**) The plane quadrangulation $\widehat{G} = \mathcal{M}(G)^*$

where we have defined $v = e^K - 1$. Now, it is obvious that for any edge-dependent factors h_e one has

$$\prod_{e \in E}(1 + h_e) = \sum_{E' \subseteq E} \prod_{e \in E'} h_e, \tag{4.5}$$

where the subset E' is defined as the set of edges for which we have taken the term h_e in the development of the product $\prod_{e \in E}$. In particular, taking $h_e = v\delta(\sigma_i, \sigma_j)$ we obtain for the partition function (4.3)

$$Z = \sum_{E' \subseteq E} v^{|E'|} \sum_{\sigma} \prod_{(ij) \in E'} \delta(\sigma_i, \sigma_j) = \sum_{E' \subseteq E} v^{|E'|} Q^{k(E')}, \tag{4.6}$$

where $k(E')$ is the number of **connected components** in the graph $G' = (V, E')$, i.e., the graph obtained from G by removing the edges in $E \setminus E'$. Those connected components are called **clusters**, and (4.6) is the **Fortuin-Kasteleyn cluster representation** of the Potts model partition function. The sum over spins σ in (4.3) has now been replaced by a sum over edge subsets, and Q appears as a parameter in (4.6) and no longer as a summation limit.

4.1.1.2 Loop Representation

We now transform the Potts model defined on a planar graph G into a model of self-avoiding loops on a related graph $\mathcal{M}(G)$, known in graph theory as the **medial graph**. Each term E' in the cluster representation (4.6) is in bijection with a term in the loop representation. The correspondence is, roughly speaking, that the loops turn around the connected components in $G' = (V, E')$ as well as their elementary internal cycles. More precisely, the loops separate the clusters from their duals.

To make this transformation precise, we first need to define the medial graph $\mathcal{M}(G) = (\tilde{V}, \tilde{E})$ carefully. Let $G = (V, E)$ be a connected planar graph with **dual** $G^* = (V^*, E^*)$. The pair (G, G^*) can be drawn in the plane such that each edge $e \in E$ intersects its corresponding dual edge $e^* \in E^*$ exactly once, see Fig. 4.1a. To each of these intersections corresponds a vertex $\tilde{i} \in \tilde{V}$ of $\mathcal{M}(G)$.

Consider now the union $G \cup G^*$. This is in fact a quadrangulation of the plane. Each quadrangle consists of a pair of half-edges and one vertex from G, and a pair

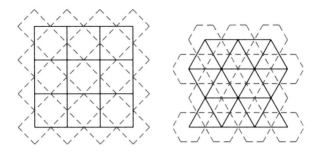

Fig. 4.2 Square and triangular lattices (*black solid lines*) with their corresponding medial lattices (*red dashed lines*)

of half-edges and one vertex from G^*. These two pairs of half-edges meet in a pair of vertices from $\tilde{V}$. An edge of $\mathcal{M}(G)$ is drawn diagonally inside each quadrangle, joining the pair of vertices from $\tilde{V}$. This defines the edge set $\tilde{E}$ and completes the definition of the medial graph. An example is shown in Fig. 4.1b.

It is manifest in these definitions that G and G^* are used in a completely symmetric way. Thus, a graph and its dual has the same medial graph, $\mathcal{M}(G) = \mathcal{M}(G^*)$. Moreover, it is easy to see that every vertex of $\mathcal{M}(G)$ has degree four.[1]

The medial of the square lattice is another (tilted) square lattice. The medial of the triangular lattice (or of its dual hexagonal lattice) is known as the **Kagomé lattice**. These two medial lattices, shown in Fig. 4.2, are particularly important for subsequent applications.

To each term E' in appearing in the sum (4.6) we now define a system of self-avoiding loops that completely cover the edges of $\mathcal{M}(G)$. Let $\tilde{i} \in \tilde{V}$ be a vertex of $\mathcal{M}(G)$ and write its adjacent (half) edges from E, E^* and $\tilde{E}$ in cyclic order as $\tilde{e}_1 e \tilde{e}_2 e^* \tilde{e}_3 e \tilde{e}_4 e^*$. Now if $e \in E'$, link up the half-edges of $\tilde{E}$ in two pairs as $(\tilde{e}_4 \tilde{e}_1)(\tilde{e}_2 \tilde{e}_3)$. Conversely, if $e \in E \setminus E'$, we link $(\tilde{e}_1 \tilde{e}_2)(\tilde{e}_3 \tilde{e}_4)$. Note that we do not allow the non-planar (crossing) linking $(\tilde{e}_1 \tilde{e}_3)(\tilde{e}_2 \tilde{e}_4)$. The set of linkings at all vertices $\tilde{V}$ defines the desired system of loops.

In concrete terms, this definition means that the loops bounce off all edges E' and cut through the corresponding dual edges. The complete correspondence is illustrated in Fig. 4.3.

To complete the transformation, note that the number of loops $l(E')$ is the sum of the number of connected components $k(E')$ and the number of independent cycles $c(E')$:

$$l(E') = k(E') + c(E'). \tag{4.7}$$

Inserting this and the topological identity

$$k(E') = |V| - |E'| + c(E') \tag{4.8}$$

---

[1] This implies that the dual of $\mathcal{M}(G)$ is a quadrangulation $\widehat{G}$, which is however different from the quadrangulation $G \cup G^*$. See Fig. 4.1c. The Potts model admits yet another representation, namely as a height model—or RSOS model—on $\widehat{G}$.

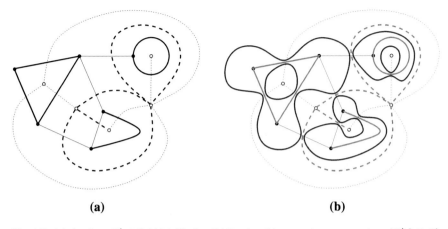

Fig. 4.3 (a) A subset $E' \subseteq E$ (*thick black solid lines*) and its complementary subset $(E')^* \subseteq E^*$ (*thick black dashed lines*). (b) The corresponding system of self-avoiding loops on the medial graph (*blue curves*)

in (4.6) we arrive at

$$Z = Q^{|V|/2} \sum_{E' \subseteq E} x^{|E'|} Q^{l(E')/2}, \tag{4.9}$$

where we have defined $x = v Q^{-1/2}$.

This is the **loop representation** of the Potts partition function. It importance stems from the fact that the loops, their local connectivities (called **linking**s in the above argument), and the non-local quantity $l(E')$ all admit an algebraic interpretation within the **Temperley-Lieb algebra**.

4.1.2 O(n) Model

The **O(n) model** is defined initially by associated with each vertex $i \in V$ of a regular planar lattice $G = (V, E)$ a vector spin $\mathbf{S}_i \in \mathbb{R}^n$ of unit length, $|\mathbf{S}_i|^2 = 1$. It turns out convenient to absorb in the integration measure a factor n/Ω_n, where Ω_n is the surface area of the unit sphere in $\mathbb{R}^n$. Thus, if S_i^α and S_i^β are components of a vector spin $\mathbf{S}_i$, we have the basic integration rule

$$\int \mathrm{d}\mathbf{S}_i \, S_i^\alpha S_i^\beta = \delta(\alpha, \beta). \tag{4.10}$$

The partition function of the O(n) model is defined by

$$Z = \int_{\mathbf{S}} \prod_{(ij) \in E} \mathrm{e}^{-V(\mathbf{S}_i, \mathbf{S}_j)}, \tag{4.11}$$

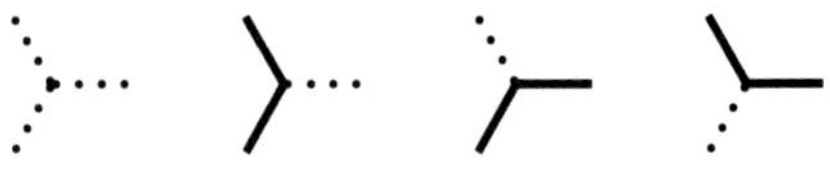

Fig. 4.4 Allowed vertices in the O(n) model on the hexagonal lattice

where we have introduced a short-hand notation for the integration over all spins

$$\int_{\mathbf{S}} := \prod_{i \in V} \left(\int d\mathbf{S}_i \right) \tag{4.12}$$

and $V(\mathbf{S}_i, \mathbf{S}_j)$ is some scalar potential describing the interaction between $\mathbf{S}_i$ and $\mathbf{S}_j$. In most texts on the O(n) model in general dimension d, one takes

$$V(\mathbf{S}_i, \mathbf{S}_j) = -K\mathbf{S}_i \cdot \mathbf{S}_j. \tag{4.13}$$

In $d = 2$, it is however much more convenient to define

$$e^{-V(\mathbf{S}_i, \mathbf{S}_j)} = 1 + K\mathbf{S}_i \cdot \mathbf{S}_j \tag{4.14}$$

where K is a dimensionless coupling constant.

The high-temperature ($K \ll 1$) expansion of (4.11) with potential (4.14) parallels the Fortuin-Kasteleyn cluster expansion of the Potts model partition function. To each term in the expansion we associate an edge subset $E' \subseteq E$, with $e = (ij) \in E$ if we take the term $K\mathbf{S}_i \cdot \mathbf{S}_j$ in (4.14). For each $i \in V$, by the symmetry $\mathbf{S}_i \to -\mathbf{S}_i$, the contribution to Z of a term associated with E' vanishes unless i is incident on an *even* number of edges in E'.

As a further simplification we now take G to be the hexagonal lattice. Since each vertex $i \in V$ has degree three, the only edge sets E' contributing to the expansion of Z are those where the vertices of $G' = (V, E')$ all have degree zero or two, as shown in Fig. 4.4. In other words, G' is a set of self-avoiding and mutually avoiding loops. The contribution to Z of a loop of length p edges is

$$Z_p = K^p \int d\mathbf{S}_1 \dots \int d\mathbf{S}_p \sum_{\alpha_1, \dots, \alpha_p} S_1^{\alpha_1} S_2^{\alpha_1} S_2^{\alpha_2} S_3^{\alpha_2} \dots S_p^{\alpha_1} S_1^{\alpha_p} \tag{4.15}$$

$$= K^p \sum_{\alpha_1, \dots, \alpha_p} \delta(\alpha_1, \alpha_2) \delta(\alpha_2, \alpha_3) \dots \delta(\alpha_p, \alpha_1) \tag{4.16}$$

$$= K^p n, \tag{4.17}$$

where we have used (4.10) repeatedly. We have then finally

$$Z = \sum_{E' \subseteq E}{}' K^{|E'|} n^{l(E')}, \tag{4.18}$$

where $l(E')$ is the number of cycles (loops), and the prime on the summation reminds us that the summation is constrained to edge subsets E' such that each vertex $i \in V$ is indicent to zero or two edges of E'.

The O(n) model partition function in the form (4.18) is quite similar to the loop representation (4.9) of the Potts model, except that the loop weight $\sqrt{Q}$ has been replaced by n, and that vertices are now allowed to be empty of loops with a relative Boltzmann weight K^{-1} proportional to the temperature.

At infinite temperature ($K = 0$), we have thus $E' = \emptyset$ and $Z = 1$. As the temperature is lowered, loops will start appearing, and one would expect that there exists some critical coupling K_c such that the average length of a loop diverges. Obviously, this means that the correlation length will diverge as well, and so K_c could be expected to be the locus of a second order phase transition. The exact solution of the $O(n)$ model however shows that these hypotheses are only fulfilled for $-2 \leq n \leq 2$. Assuming this to be the case, if K_c is small enough, one could hope that the critical behaviour is identical to that of the generic $O(n)$ model, since the two potentials (4.13)–(4.14) agree to first order in K. The exact solution implies that one has in fact

$$K_c = (2 \pm \sqrt{2 - n})^{-1/2}. \tag{4.19}$$

4.1.3 Vertex and Height Models

In the definition of the Q-state Potts and the $O(n)$ models, the parameters Q and n were originally positive integers. However, in the corresponding loop models, (4.9) and (4.18), they appear as formal parameters and may thus take arbitrary complex values. The price to pay for this generalisation is the appearance of a non-locally defined quantity, the **number of loops** l. The locality of the models may be recovered by transforming them to vertex models with complex Boltzmann weights as we now show.

4.1.3.1 From Loops to Arrows

The following argument supposes that $G = (V, E)$ is a (connected) planar graph. Most applications however suppose a regular lattice, a situation to which we shall return shortly.

Consider a model of self-avoiding loops defined on G (or some related graph, such as the medial graph $\mathcal{M}(G)$ for the Potts model). The Boltzmann weights are supposed to consist of a local piece—depending on if and how the loops pass through a given vertex—and a non-local piece of the form n^l, where n is the loop weight and l is the number of loops. In the case of the Potts model we have $n = \sqrt{Q}$.

In a first step, each loop is independently decorated by a global orientation $s = \pm 1$, which by planarity and self-avoidance can be described as either counterclockwise ($s = 1$) or clockwise ($s = -1$). If each oriented loop is given a weight $w(s)$, we have the requirement

$$n = w(1) + w(-1). \tag{4.20}$$

An obvious possibility, sometimes referred to as the **real loop ensemble**, is $w(1) = w(-1) = n/2$. This can be interpreted as an $O(n/2)$ model of complex spins.

We are however more interested in the **complex loop ensemble** with $w(s) = e^{is\gamma}$.
Note that in the expected critical regime,

$$n = 2\cos\gamma \in [-2, 2], \tag{4.21}$$

the parameter $\gamma \in [0, \pi]$ is real. *Locality* is retrieved by remarking that the weights
$w(\pm 1)$ are equivalent to assigning a local weight $w(\alpha/2\pi)$ each time the loop turns
an angle α (counted positive for left turns).

Note that a planar graph cannot necessarily be drawn in the plane in such a way
that all edges are straight line segments. Therefore, the local weights $w(\alpha/2\pi)$ must
in general be assigned both to vertices *and* to edges. However, it is certainly possible to redraw the graph so that each edge is a succession of several straight line
segments. Introducing auxiliary vertices of degree two at the places where two segments join up, the weight for turning can be assigned to those auxiliary vertices. In
that sense, any planar graph admits a local redistribution of the loop weight, with
local weights $w(\alpha/2\pi)$ assigned only to vertices.

The loop model is now transformed into a **local vertex model** by assigning to
each edge traversed by a loop the orientation of that loop. An edge not traversed by
any loop is assigned no orientation. The total vertex weight is determined from the
configuration of its incident oriented edges: it equals the above local loop weights
summed over the possible linkings of oriented loops which are compatible with the
given edge orientations. In addition, one must multiply this by any loop-independent
local weights, such as x in (4.9) or K in (4.18).

To see how this is done, it most convenient to turn to some representative examples on the regular square lattice.

4.1.3.2 Six-Vertex Model

Consider first the Potts model on the square lattice G. The loop model is defined on
the corresponding medial lattice $\mathcal{M}(G)$ which is another (tilted) square lattice. Each
edge of the lattice is visited by a loop, and two loop segments (possibly parts of the
same loop) meet at each vertex. In the oriented loop representation, each vertex is
therefore incident on two outgoing and two in-going edges—in other words, it has
in-degree two and out-degree two.

It is convenient for the subsequent discussion to make the couplings of the Potts
model anisotropic. In its original spin formulation (4.3) we therefore let K_1 (resp.
K_2) denote respectively the dimensionless coupling in the horizontal (resp. vertical)
direction of the square lattice, and we let

$$x_1 = \frac{e^{K_1} - 1}{\sqrt{Q}}, \qquad x_2 = \frac{e^{K_2} - 1}{\sqrt{Q}} \tag{4.22}$$

be the corresponding parameters appearing in the loop representation (4.9). Note
that in all the results obtained this far it is straightforward to generalise to completely
inhomogeneous edge dependent couplings, and the only reason that we have chosen
not to present the results in this generality is that it tends to make notations slightly
cumbersome.

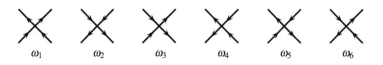

Fig. 4.5 Weights in the six-vertex model

The six possible configurations of arrows around a vertex of the medial lattice $\mathcal{M}(G)$ are shown in Fig. 4.5. The corresponding vertex weights are denoted ω_p (resp. ω'_p) on the even (resp. odd) sub-lattice of $\mathcal{M}(G)$. By definition, a **vertex** of the even (resp. odd) sub-lattice of $\mathcal{M}(G)$ is the mid point of an edge with coupling K_1 (resp. K_2) of the original spin lattice G. With respect to Fig. 4.5 we define the **even sub-lattice** to be such that an edge $e \in E$ is horizontal, and the corresponding dual edge $e^* \in E^*$ is vertical. For the **odd sub-lattice**, exchange e and e^*.

Using (4.9) we then have

$$Z = Q^{|V|/2} \sum_{\text{arrows}} \prod_{p=1}^{6} (\omega_p)^{N_p} (\omega'_p)^{N'_p}, \qquad (4.23)$$

where the sum is over arrow configurations satisfying the constraint "two in, two out" at each vertex, and N_p (resp. N'_p) is the number of vertices on the even (resp. odd) sub-lattice with arrow configuration p. This is the **six-vertex representation** of the square-lattice Potts model. The weights read explicitly

$$\omega_1, \ldots, \omega_6 = 1, 1, x_1, x_1, e^{i\gamma/2} + x_1 e^{-i\gamma/2}, e^{-i\gamma/2} + x_1 e^{i\gamma/2} \qquad (4.24)$$

$$\omega'_1, \ldots, \omega'_6 = x_2, x_2, 1, 1, e^{-i\gamma/2} + x_2 e^{i\gamma/2}, e^{i\gamma/2} + x_2 e^{-i\gamma/2}. \qquad (4.25)$$

To see this, note that configurations $i = 1, 2, 3, 4$ are compatible with just one linking of the oriented loops:

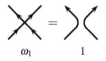

whereas $i = 5, 6$ are compatible with two different linkings (and the weight is obtained by summing over these two):

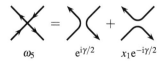

Note that the even and odd sub-lattices are related by a $\pi/2$ rotation of the vertices in Fig. 4.5. This rotation interchanges configurations $(\omega_1, \omega_2) \leftrightarrow (\omega'_3, \omega'_4)$ and $\omega_5 \leftrightarrow \omega'_6$. On the level of the weights it corresponds to $x_1 \leftrightarrow x_2$.

4.1.3.3 From Arrows to Heights

The vertex models can be turned into **height model**s, often referred to as **solid-on-solid model**s, or *SOS models* for short. To this end, assign a scalar variable $h(f)$ to each lattice face f (i.e., to each vertex of the lattice *dual* to the one on which the loop model has been defined), so that h increases (resp. decreases) by a each time one traverses a left-going (resp. right-going) edge. This definition of the height h is consistent, since each vertex is incident on as many in-going as outgoing edges. Since this defines only height differences, one completes the definition of h by arbitrarily fixing $h(f_0) = 0$, where f_0 is some fixed face.

The choice of the elementary height jump a is completely arbitrary and subsequent physical results are independent of it. It is conventional to set $a = \pi$ for the Potts and $O(n)$ models.

In the arrow formulation the interaction was among the arrows incident on a common vertex. By duality, this turns into an interaction between the heights around a common face of the dual lattice. For that reason, the height models are also known as *interaction-round-a-face models*, or **IRF model**s for short.

We should also point out that when dealing with loop models which are more involved than those presented in the present section, it may be necessary to take the height variable as vector valued.

4.1.3.4 Twisted Vertex Model

Sometimes it is convenient to consider particular correlation functions in which the weight of some of the loops are changed. As an elementary example, consider the Potts loop model defined on a connected planar graph $G = (V, E)$ and let $i_1, i_2 \in V$ be a pair of root vertices.[2] The partition function $Z(n)$ is given by (4.9) with loop weight $n = \sqrt{Q}$ and additional local weights at the vertices.

Define now a modified partition function $Z_1(n, n_1)$ as follows: loops on $\mathcal{M}(G)$ surrounding neither or both of the roots have an unchanged weight n, whereas those surrounding only one of the roots have a modified weight n_1. This defines the two-point correlation function $Z_1(n, n_1)/Z(n)$. An interesting special case is provided by $n_1 = 0$, which expresses the probability that the two roots belong to the same cluster.

It is possible to produce $Z_1(n, n_1)$ in the vertex model representation, leading to a so-called **twisted vertex model**. To this end, let $\mathcal{P}_{12}$ be an oriented self-avoiding path on G, going from i_1 to i_2. Let us parametrise

$$n_1 = 2\cos\gamma_1 \in [-2, 2] \tag{4.26}$$

with real $\gamma_1 \in [0, \pi]$. In the arrow formulation, we then associate a special weight $\tilde{w}$ to any edge $\tilde{e}$ of $\mathcal{M}(G)$ that crosses the path $\mathcal{P}_{12}$. The weight $\tilde{w}$ depends on

---

[2]The same construction can be taken over for the $O(n)$ loop model (4.18) on condition that the roots be located on the dual lattice, and by allowing for other obvious modifications.

the orientation of the arrow on $\tilde{e}$: it equals $e^{i\gamma_1}$ (resp. $e^{-i\gamma_1}$) if the arrow points from left to right (resp. from right to left) upon viewing $\tilde{e}$ along the direction given by $\mathscr{P}_{12}$.

The path $\mathscr{P}_{12}$ is often called a **seam**, and the edges traversing it are referred to as **seam edge**s.

In the oriented loop representation, it is easy to see that a loop surrounding neither or both of the roots will traverse $\mathscr{P}_{12}$ an even number of times, and the phase factors $\tilde{w}$ will cancel out globally. However, a loop surrounding just one of the roots with have one excess factor $e^{\pm i\gamma_1}$ depending on its global orientation (clockwise or counterclockwise), leading to (4.26) once the orientations have been summed over.

Note that the above construction of $Z_1(n, n_1)$ depends on the seam $\mathscr{P}_{12}$ only through its end points i_1 and i_2. In that sense, the exact shape of the seam is irrelevant and can be deformed at will.

Finally, the weights $\tilde{w}$ can be absorbed in the vertex weights, by incorporating them in the weight of the vertex at the right (with respect to the orientation defined by $\mathscr{P}_{12}$) end point of $\tilde{e}$.

4.2 Coulomb Gas in the Bulk

It has been known since the 1970's that the critical point of many two-dimensional models of statistical physics can be identified with a Gaussian free-field theory. A general framework for the computation of critical exponents was first given in 1977 by José et al. in the so-called spin wave picture. This was further elaborated in the early 1980's by den Nijs and Nienhuis into what has become known as the *Coulomb gas* (CG) construction.

The CG approach is particularly suited to deal with the continuum limit of lattice models of closed loops, in which each loop carries a Boltzmann weight n.

The marriage between the CG and conformal field theory (CFT) happened in 1986–87, when Di Francesco, Saleur and Zuber made the loop model $\leftrightarrow$ CG correspondence more precise and showed how the ideas of modular invariance can be put to good use in the study of loop models. At the same time, Duplantier and Saleur developed a range of applications to polymer physics.

We have seen how the loop models can be transformed into height models with local (albeit complex) Boltzmann weights. It is the continuum limit of this height which acts as the conformally invariant free field. The underlying lattice model implies that this height field is compactified. The naive free-field action however needs to be modified with extra terms, traditionally known as background and screening electric charges, see also chapter 1. The geometrical significance of these terms has been greatly clarified by Kondev and collaborators. The resulting CFT, known as a *Liouville field-theory*, will be written down in this section.

4.2.1 Compactified Free Boson

In the continuum limit, we expect the local height field h to converge to a free bosonic field $\phi(\mathbf{r})$, whose entropic fluctuations are described by an action of the form

$$S \sim g \int d^2\mathbf{r} \, (\nabla\phi)^2, \tag{4.27}$$

with coupling $g = g(n)$ which is a monotonically increasing function of n. In particular, for $n \to \infty$ the lattice model is dominated by the configuration where loops of the minimal possible length cover the lattice densely; the height field is then flat, $\phi(\mathbf{r}) = $ constant, and the correlation length ξ is of the order of the lattice spacing. For finite but large n, ϕ will start fluctuating, loop lengths will be exponentially distributed, and ξ will be of the order of the linear size of the largest loop. When $n \to n_c^+$, for some critical n_c (we shall see that $n_c = 2$), this size will diverge, and for $n \le n_c$ the loop model will be conformally invariant with critical exponents that depend on $g(n)$. The interface described by $\phi(\mathbf{r})$ is then in a **rough phase**.

The remainder of this section is devoted to making this intuitive picture more precise, and to refine the free bosonic description of the critical phase.

As a first step towards greater precision, we now argue that $\phi(\mathbf{r})$ is in fact a **compactified boson**. To see this, it is convenient to consider the *oriented* loop configurations that give rise to a maximally flat microscopic height h; we shall refer to them as **ideal states**.

For the Potts model on the square lattice, an ideal state is a dense packing of length-four loops, all having the same orientation. There are four such states, corresponding to two choices of orientation and two choices of the sub-lattice of lattice faces surrounded by the loops. An ideal state can be gradually changed into another by means of $\sim N$ local changes of the transition system and/or the edge orientations. As a result, the mean height will change, $\phi \to \phi \pm a$. Iterating this, one sees that one may return to the initial ideal state whilst having $\phi \to \phi \pm 2a$. For consistency, we must therefore require $\phi(\mathbf{r}) \in \mathbb{R}/(2a\mathbb{Z})$, i.e., the field is compactified indeed.

The same construction, applied to the O(n) model, yields six ideal states of oriented length-six loops (resulting from a choice of three sub-lattices and two orientations). Changing the ideal state in four steps, as shown in Fig. 4.6, produces the initial state but with a height change of $\pm 2a$. So one has the same compactification radius, $\phi(\mathbf{r}) \in \mathbb{R}/(2a\mathbb{Z})$, as in the case of the Potts model. We shall follow standard conventions and set $a = \pi$ in what follows.

Note that we may already suspect—and we shall see below in more detail— that *the O(n) model in the dense phase and the Q-state Potts model give identical critical theories in the continuum limit, for $n = \sqrt{Q}$*. However, the correspondence between operators in the microscopic model and the continuum limit is not necessarily identical, leading to subtle differences. For instance, the energy operators of the two models become different objects in the continuum limit.

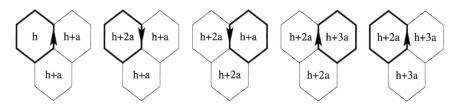

Fig. 4.6 Ideal states of the O(n) model on the hexagonal lattice. For each of the *five panels*, the state of the complete infinite lattice is obtained by tiling the plane with the three faces shown, while respecting the three-sub-lattice structure. The different *panels* are related, from *left* to *right*, by the construction explained in the main text, under which one ideal state is gradually changed into another. The *leftmost* and *rightmost panels* represent the same ideal state, but with a global height change $\phi \rightarrow \phi + 2a$ that determines the compactification radius [15]

4.2.2 Liouville Field-Theory

The essence of the above discussion is that the critical properties of the loop models under consideration can be described by a continuum-limit partition function that takes the form of a functional integral

$$Z = \int \mathscr{D}\phi(\mathbf{r}) \exp\left(-S[\phi(\mathbf{r})]\right). \tag{4.28}$$

Here $S[\phi(\mathbf{r})]$ is the Euclidean action of the compactified scalar field $\phi(\mathbf{r}) \in \mathbb{R}/(2\pi\mathbb{Z})$. The hypothesis that the critical phase is described by bounded elastic fluctuations around the ideal states means that S must contain a term

$$S_{\mathrm{E}} = \frac{g}{4\pi} \int \mathrm{d}^2\mathbf{r}\,(\nabla\phi)^2 \tag{4.29}$$

with coupling constant $g > 0$. Higher derivative terms that one may think of adding to (4.29) can be ruled out by the $\phi \rightarrow -\phi$ symmetry, or by arguing *a posteriori* that they are RG-irrelevant in the full field theory that we are about to construct.

Note that the partition function (4.28) does not purport to coincide with (4.3) or (4.18) on the scale of the lattice constant. (A similar remark holds true for the correlation functions that one may similarly write down.) We do however claim that their long-distance properties are the same. In that sense, the CG approach is an exact, albeit by no means rigorous, method for computing critical exponents and related quantities. A more precise equivalence between discrete and continuum-limit partition functions can however be achieved on a torus.

The action (4.29) is that of a compactified boson. To obtain the full physics of the loop model, one needs however, to add two more terms to the action, as we now shall see.

To proceed, we consider the underlying lattice model as being defined on a cylinder, $\mathbf{r} = (r, t)$. This has the advantage of making direct contact with the radial quantisation formalism. The boundary conditions are thus periodic in the space direction, $r = r + L$, and free in the time (t) direction. Ultimately, the results obtained on the cylinder can always be transformed into other geometries by means of a conformal mapping.

With this geometry, the equivalence between the loop model and a local height model with complex weights must be revisited. One proceeds as in our derivation of the twisted vertex model, taking the two root vertices to reside on opposite ends of the cylinder.

While loops homotopic to a point still acquire their correct global weight n from the local angle-dependent weights $w(\alpha/2\pi)$, this is no longer true for loops that wind around the cylinder. Summing over loop orientations, their weight would be $\bar{n} = 1 + 1 = 2$. Consider now adding a term

$$S_B = \frac{ie_0}{4\pi} \int d^2\mathbf{r}\, \phi(\mathbf{r})\mathscr{R}(\mathbf{r}) \tag{4.30}$$

to the effective action S, where $\mathscr{R}$ is the scalar curvature[3] of the space $\mathbf{r}$. The parameter e_0 is known in CG language as the **background electric charge**. On the cylinder, one has simply $S_B = ie_0(\phi(r, \infty) - \phi(r, -\infty))$, meaning that in the partition function (4.28) an oriented loop with winding number $q = 0, \pm 1$ (all other winding numbers are forbidden by the self-avoidance of the loops) can equivalently be assigned an extra weight of $\exp(i\pi q e_0)$.

For non-winding loops ($q = 0$) this does not change the reasoning of the height mapping, whilst summing over the two orientations ($q = \pm 1$) of a winding loop produces the weight $\bar{n} = 2\cos(\pi e_0)$. The choice $e_0 = \gamma/\pi$ will thus assign to a winding loop the same weight $\bar{n} = n$ [see (4.21)] as to a non-winding one (but note that other choices leading to $\bar{n} \neq n$ may be useful in some applications of the CG technique).

The object $e^{ie\phi}$ (or more precisely, its normal ordered product $:e^{ie\phi}:$) is known in field theory as a **vertex operator** of (electric) charge e. The boundary term (4.30) thus corresponds to the insertion of two oppositely charged vertex operators at either end of the cylinder.

At this stage two problems remain: the field theory does not yet take account of the weight n of contractible loops, and the coupling constant g has not yet been determined. These two problems are closely linked, and allow us to fix exactly $g = g(n)$. The idea is to add a further **Liouville term**

$$S_L = \int d^2\mathbf{r}\, w[\phi(\mathbf{r})] \tag{4.31}$$

to the action, which then reads in full

$$S[\phi(\mathbf{r})] = S_E + S_B + S_L. \tag{4.32}$$

In (4.31), $e^{-w[\phi(\mathbf{r})]}$ is the scaling limit of the microscopic vertex weights w_i. To identify it we show the argument for the O(n) model, the Potts case being similar.

Due to the compactification, $S_L[\phi]$ is a periodic functional of the field, and as such it can be developed as a Fourier sum over vertex operators

$$w[\phi] = \sum_{e \in \mathscr{L}_w} \tilde{w}_e\, e^{ie\phi}, \tag{4.33}$$

---

[3]We consider the scalar curvature in a generalised sense, so that delta function contributions may be located at the boundaries. Implicitly, we are just applying the Gauss-Bonnet theorem.

where $\mathcal{L}_w$ is some sub-lattice of $\mathcal{L}_0 = \mathbb{Z}$. Note that $\mathcal{L}_w$ may be a proper sub-lattice of $\mathcal{L}_0$ if $w[\phi]$ has a higher periodicity than that trivially conferred by the compact-ification of ϕ. By inspecting Fig. 4.6, we see that this is indeed the case here: the (geometric) averages of the microscopic weights coincide on the first, third, and fifth panels, indicating that the correct choice is $\mathcal{L}_w = 2\mathcal{L}_0$. This intuitive deriva-tion of $\mathcal{L}_w$ (which can easily be corroborated by considering more complicated microscopic configurations) demonstrates the utility of the ideal state construction.

We recall some important properties of the compactified boson with action S_E, see also chapter 1. Its central charge is $c = 1$ and the scaling dimension $x_{e,m}$ of an operator (see Eq. (1.10)) with electromagnetic charge (e, m) is given by

$$x_{e,m} = 2\Delta_{e,m} = \frac{1}{2}\left[\frac{e^2}{g} + gm^2\right].$$
(4.34)

Having now identified the electric charge e with that of the vertex operator $e^{ie\phi}$, one could alternatively readily re-derive (4.34) by computing the two-point function $\langle e^{ie\phi(\mathbf{r})}e^{-ie\phi(\mathbf{r}')}\rangle$ by standard Gaussian integration.

The magnetic charge m corresponds to dislocations in the height field ϕ due to the presence of defect lines. Below we shall see how to identify these defect lines with the extremities of polymers and compute the related critical exponents.

It remains to assess how the properties of the compactified boson are modified by the inclusion of the term S_B. Physical reasoning consists in arguing that the vertex operators $e^{\pm ie_0\phi}$ will create a "floating" electric charge of magnitude $2e_0$ that "screens" that of the other fields in any given correlation function. We infer that (4.34) must be changed into

$$x_{e,m} = 2\Delta_{e,m} = \frac{1}{2}\left[\frac{e(e - 2e_0)}{g} + gm^2\right].$$
(4.35)

Note that our normalisation is such that both e and m are integer.

4.2.3 Marginality Requirement

Following Kondev we now claim that *the Liouville potential S_L must be exactly marginal*. This follows from the fact that all loops carry the same weight n, inde-pendently of their size, and so the term S_L in the action that enforces the loop weight must not renormalise under a scale transformation. The most relevant vertex opera-tor appearing in (4.33) has charge $e_w = 2\pi/a = 2$, and so $x_{e_w,0} = 2$. Using (4.35), this fixes the coupling constant as $g = 1 - e_0$. In other words, the loop weight has been related to the CG coupling as

$$n = \pm\sqrt{Q} = -2\cos(\pi g)$$
(4.36)

with $0 < g \leq 1$ for the Potts model or the dense O(n) model.

The term S_B shifts the ground state energy with respect to the $c = 1$ theory described by S_E alone. The corrected central charge is then $c = 1 + 12x_{e_0,0}$. This gives

$$c = 1 - \frac{6(1-g)^2}{g}. \tag{4.37}$$

It should be noted that the choice $e_w = 2$ is not the only one possible. Namely, the coefficient $\tilde{w}_{e_w}$ of the corresponding vertex operator in (4.33) may be made to vanish, either by tuning the temperature T in the O(n) model, or by introducing non-magnetic vacancies in the Potts model. The former case corresponds to taking the high-temperature solution [plus sign] in (4.19), while the latter amounts to being at the tricritical point of the Potts model. The next-most relevant choice is then $\tilde{e}_w = -2$, and going through the same steps as above we see that one can simply maintain (4.36), but take the coupling in the interval $1 \leq g \leq 2$ for the dilute O(n) model or the tricritical Potts model.

The electric charge e_w whose vertex operator is required to be exactly marginal is known as the **screening charge** in standard CG terminology.

The central charge (4.37) can now be formally identified with that of the Kac table for unitary minimal models

$$c = 1 - \frac{6}{m(m+1)}, \tag{4.38}$$

$$\Delta_{r,s} = \frac{((m+1)r - ms)^2 - 1}{4m(m+1)}. \tag{4.39}$$

The result is a formal relation between the minimal model index m and the CG coupling g, valid for integer m. We have

$$m = \begin{cases} \frac{g}{1-g} & \text{for the dense O(n) model, or the critical Potts model} \\ \frac{1}{g-1} & \text{for the dilute O(n) model.} \end{cases} \tag{4.40}$$

Note that this identification holds also for non-integer m, when the loop models are non-unitary. Even in that case, critical exponents can be conveniently written in terms of the conformal weights $\Delta_{r,s}$ and often correspond to integer choices of r, s.

The special cases $n \to 0$ are related to self-avoiding walks and polygons. This gives $g = 1/2$ for dense polymers (with $c = -2$ and $m = 1$), and $g = 3/2$ for dilute ones (with $c = 0$ and $m = 2$).

4.2.4 Critical Exponents

The Coulomb gas technology can be used to compute a variety of critical exponents in loop models, and in the related Potts and O(n) models. Rather than insisting on completeness, we shall treat a simple and significant example of the so-called *watermelon exponents*.

An important object in loop models is the operator $\mathcal{O}_\ell(\mathbf{r}_1)$ that inserts ℓ oriented lines at a given point $\mathbf{r}_1$. Microscopically, this can be achieved by violating the arrow

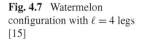

Fig. 4.7 Watermelon configuration with $\ell = 4$ legs [15]

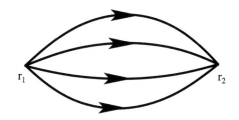

conservation constraint at $\mathbf{r}_1$. For instance, in the $O(n)$ model one can allow a vertex which is adjacent to one outgoing and two empty edges. Doing so at ℓ vertices in a small region around $\mathbf{r}_1$ yields a microscopic realisation of the composite operator $\mathscr{O}_\ell(\mathbf{r}_1)$.

If one had strict arrow conservation at all other vertices, the insertion of $\mathscr{O}_\ell(\mathbf{r}_1)$ would not lead to a consistent configuration. However, also inserting $\mathscr{O}_{-\ell}(\mathbf{r}_2)$, the operator that absorbs ℓ oriented lines in a small region around $\mathbf{y}$, will lead to consistent configurations (see Fig. 4.7) in which ℓ defect lines propagate from $\mathbf{r}_1$ to $\mathbf{r}_2$. Let $Z_\ell(\mathbf{r}_1, \mathbf{r}_2)$ be the corresponding constrained partition function. One then expects

$$\langle \mathscr{O}_\ell(\mathbf{r}_1)\mathscr{O}_{-\ell}(\mathbf{r}_2)\rangle := \frac{Z_\ell(\mathbf{r}_1, \mathbf{r}_2)}{Z} \sim \frac{1}{|\mathbf{r}_1 - \mathbf{r}_2|^{2x_\ell}} \quad \text{for } |\mathbf{r}_1 - \mathbf{r}_2| \gg 1. \quad (4.41)$$

The corresponding critical exponents $x_\ell = \Delta_\ell + \bar{\Delta}_\ell$ are known as **watermelon** (or *fuseau*, or ℓ-leg) **exponents**. To compute them, we first notice that the sum of the height differences around a closed contour encircling $\mathbf{r}_1$ but not $\mathbf{r}_2$ will be $a\ell$. Equivalently, one could place the two defects at the extremities of a cylinder [i.e., taking $\mathbf{r}_1 = (r, -\infty)$ and $\mathbf{r}_2 = (r, \infty)$], and the height difference would be picked up by any non-contractible loop separating $\mathbf{r}_1$ and $\mathbf{r}_2$.

A little care is needed to interpret the configurations of $Z_\ell(\mathbf{r}_1, \mathbf{r}_2)$ in the model of un-oriented loops. The fact that all ℓ lines are oriented away from $\mathbf{r}_1$ prevents them from annihilating at any other vertex than $\mathbf{r}_2$. One should therefore like to think about them as ℓ marked lines linking $\mathbf{r}_1$ and $\mathbf{r}_2$, where each line carries the Boltzmann weight 1. This is consistent with not summing over the orientations of the defect lines in the oriented loop model.

However, each oriented line can also pick up spurious phase factors $w(\alpha/2\pi)$, due to the local redistribution of loop weights, whenever it turns around the end points $\mathbf{r}_1$ and $\mathbf{r}_2$. These factors are however exactly cancelled if we insert in addition a vertex operator $e^{ie_0\phi}$ (resp. $e^{-ie_0\phi}$) at $\mathbf{r}_1$ (resp. $\mathbf{r}_2$). The argument is exactly the same as the one used to motivate the background charge e_0 in the first place. Note that these vertex operators do not modify the weighting of closed loops, since these must encircle either none or both of $\mathbf{r}_1, \mathbf{r}_2$. We conclude that $x_\ell = x_{e_0, m_\ell}$, and using (4.35) this gives

$$x_\ell = \frac{1}{8}g\ell^2 - \frac{(1-g)^2}{2g}. \quad (4.42)$$

Interestingly, these exponents can be attributed to the Kac table under the identification (4.40). One has

$$x_\ell = \begin{cases} 2\Delta_{0,\ell/2} & \text{for the dense } O(n) \text{ model} \\ 2\Delta_{\ell/2,0} & \text{for the dilute } O(n) \text{ model.} \end{cases} \tag{4.43}$$

The appearance of half-integer indices is somewhat puzzling, whereas the fact that these exponents are located outside the fundamental domain of the Kac table reflects the non-local nature of the watermelon operators.

It should be noticed that x_4 is irrelevant (resp. relevant) in the dilute (resp. dense) phase of the $O(n)$ model, i.e., for $1 < g < 2$ (resp. $0 < g < 1$). This means that on lattices with vertices of degree ≥ 4, loop self-intersections are irrelevant in the dilute phase. On the other hand, for the dense phase such self-intersections are relevant and will induce a flow to a supersymmetric Goldstone phase that is not described by the CG approach. In other words, Nienhuis' original approximation of the true $O(n)$ model that led to (4.18) is exact in the continuum limit, but only in the dilute phase.

4.2.4.1 Application to Percolation Clusters

The watermelon exponents can be used to elucidate the fractal properties of the Fortuin-Kasteleyn (FK) clusters defined in Sect. 4.1.1.1. Here we limit the discussion to the special case of percolation clusters.

Bond percolation is the $Q \to 1$ limit of FK clusters. We have therefore $g = \frac{2}{3}$ from (4.36). The watermelon exponents (4.42) are thus

$$x_\ell = \frac{\ell^2 - 1}{12}. \tag{4.44}$$

Marking a point $\mathbf{r}$ on the hull of a percolation cluster corresponds to the insertion of the operator $\mathcal{O}_2(\mathbf{r})$. Indeed, since the cluster is supposed to be infinite, the two pieces of the hull that go away from the marked point will effectively persist all the way to "infinity"; they hence behave just like two self-avoiding legs. The **fractal dimension** of the hull is therefore

$$d_{\mathrm{h}} = 2 - x_2 = \frac{7}{4}. \tag{4.45}$$

A **pivotal edge** is defined as an edge belonging to a percolation cluster which is such that the removal of the edge makes the cluster break into two connected components. In the literature on percolation, pivotal edges are also known as **red bonds**. Cutting the loop strands on either side of any edge belonging to the cluster looks like an $\ell = 4$ leg insertion. Note however that only if the edge is pivotal will the four legs propagate to "infinity" without contracting among themselves. Therefore the fractal dimension of red bonds is

$$d_{\mathrm{rb}} = 2 - x_4 = \frac{3}{4}. \tag{4.46}$$

4.2.4.2 Magnetic Exponent

The watermelon exponents can be said to be of the "magnetic" type, since they induce a magnetic type defect charge m_ℓ in the CG. The standard magnetic exponent, describing the decay of the spin-spin correlation function in the Potts model, is however not of the watermelon type. It can nevertheless be inferred from (4.35) as follows:

The probability that two spins situated at $\mathbf{r}_1$ and $\mathbf{r}_2$ are in the same Potts state is proportional, in the random cluster picture, to the probability that they belong to the same cluster. In the cylinder geometry this means that no winding loop separates $\mathbf{r}_1$ from $\mathbf{r}_2$. This can be attained in the CG by giving a weight $n_1 = 0$ to such loops. We have seen that inserting a pair of vertex operators with charge $\pm e$ at $\mathbf{r}_1$ and $\mathbf{r}_2$ leads exactly to this situation with $n_1 = 2\cos(\pi e)$, and so we need $e = \frac{1}{2}$. The scaling dimension of this excitation, with respect to the ground state which has $e = e_0$, is then

$$x_m = x_{\frac{1}{2},0} - x_{e_0,0} = \frac{1 - 4(1-g)^2}{8g}. \tag{4.47}$$

In particular for the Ising model, with $g = \frac{3}{4}$, this yields the magnetic exponent $x_\sigma := x_m = \frac{1}{8}$, or in standard notation

$$\beta = \frac{1}{8} \quad \text{(Ising model)}. \tag{4.48}$$

For bond percolation, with $g = \frac{2}{3}$, we find $x_m = \frac{5}{48}$. The fractal dimension of a percolation cluster is thus

$$d_c = 2 - x_m = \frac{91}{48}. \tag{4.49}$$

The location in the Kac table (4.39) of the magnetic exponent (4.47) can be found using (4.40):

$$x_m = 2\Delta_{1/2,0}. \tag{4.50}$$

Note that this differs from the lowest possible watermelon excitation $x_2 = 2\Delta_{0,1}$. Indeed, the two-leg excitation corresponds to a cluster that propagates along the length direction of the cylinder *without* wrapping around the transverse periodic direction. The dominant configurations participating in the magnetic correlation function have *no* propagating legs, since the cluster containing $\mathbf{r}_1$ and $\mathbf{r}_2$ will typically wrap around the cylinder.

4.3 Coulomb Gas at the Boundary

The aspects of CFT exposed to this point pertain to unbounded geometries, either that of the infinite plane (Riemann sphere) or that of the torus (which is really a finite geometry made unbounded through the periodic boundary conditions). In contrast,

boundary conformal field-theory (BCFT) describes surface critical behaviour, i.e., a critical system confined to a bounded geometry. The simplest such geometry, and probably the most relevant from the point of view of polymer physics, is that of the upper half plane $\{z \mid \operatorname{Im} z \geq 0\}$, where the real axis $\mathbb{R}$ acts as the boundary (one-dimensional "surface").

The foundations of BCFT were laid by Cardy who also initiated many of the subsequent developments and applications.

4.3.1 Surface Critical Behaviour

To convey an idea of which phase transitions may result from the interplay between bulk and boundary degrees of freedom, and what may be the corresponding boundary conditions, we begin by a qualitative discussion of a simple magnetic spin system, which is strictly valid only for $d > 2$ dimensions. We denote the local order parameter (magnetisation) by ϕ. When the boundary spins enjoy free boundary conditions, they interact more weakly than the bulk spins, since microscopically they are coupled to fewer neighbouring spins. Upon lowering the temperature, the bulk will therefore order before the surface: this is the so-called **ordinary transition**. Now consider placing the system slightly below the bulk critical temperature. Then ϕ is non-zero deep inside the bulk, and will decrease upon approaching the boundary. One can argue that in the continuum limit ϕ will vanish exactly on the boundary. Thus, the Dirichlet boundary condition $\phi|_{\mathbb{R}} = 0$ is the appropriate choice for describing the ordinary transition.

Let us now introduce a coupling J_s between nearest-neighbour spins on the boundary which may be different from the usual bulk coupling constant J. Taking $J_s > J$ one may "help" the boundary to order more easily.[4] When J_s takes a certain critical value, we are at the **special transition**, at which the bulk and the boundary order simultaneously. Finally, when $J_s \to \infty$ the boundary spins are always completely ordered,[5] a fact which changes the nature of the ordering transition of the bulk, now referred to as the **extraordinary transition**. This corresponds to the Dirichlet boundary condition $\phi|_{\mathbb{R}} = \infty$ in the continuum limit. Note that in the application of boundary CFT to loop models the meaning of J_s is to give a specific fugacity to monomers on the boundary.

The control parameter J_s can be thought of in a renormalisation group sense, and is readily seen to be irrelevant at the ordinary and extraordinary transitions. Accordingly we expect a boundary RG flow to go from the special to either of the two other transitions. (In the case of the $2D$ Ising model, the special and extraordinary transitions actually coincide.)

---

[4]A similar effect could be obtained by adding a surface magnetic field, but here we do not wish to break the symmetry of the model [typically O(n) in applications to loop models].

[5]This should not (as is sometimes seen in the literature) be confused with imposing *fixed* boundary conditions, which would rather correspond to an infinite symmetry-breaking field applied on the boundary (and is sometimes referred to as **normal transition**).

In our subsequent application to loop models, we rather think of ϕ as a height field which is dual to the system of oriented loops (this is the so-called Coulomb gas approach). In other words, the loops are level lines of ϕ. Dirichlet boundary conditions then describe a situation in which loops are reflected off the boundary, and adjoining two different Dirichlet conditions forces one or more "loop ends" to emanate from the boundary. One may also impose Neumann boundary conditions, $\partial \phi / \partial y|_{\mathbb{R}} = 0$, meaning that the "loops" coming close to the boundary must in fact terminate perpendicular to it. Clearly the non-local aspects of these situations call for a more detailed discussion.

4.3.2 Sketch of Boundary CFT

We assume known the principal results of bulk CFT and focus here the discussion on the main differences with the boundary case (see also Chap. 1).

The allowed conformal mappings in BCFT must keep invariant both the boundary itself and the boundary conditions imposed along it. For the global conformal transformations

$$w(z) = \frac{a_{11}z + a_{12}}{a_{21}z + a_{22}} \tag{4.51}$$

the invariance of the real axis forces $a_{ij} \in \mathbb{R}$, i.e., they form the group $Sl(2, \mathbb{R})$ and the number of parameters is halved from 6 to 3. For an infinitesimal local conformal transformation $z \rightarrow w(z) = z + \varepsilon(z)$ the requirement reads $\varepsilon(\bar{z}) = \bar{\varepsilon}(z)$. This property can be used to eliminate the $\bar{\varepsilon}(z)$ part altogether, since it is just the analytic continuation of $\varepsilon(\bar{z})$ into the lower half plane. It follows that $\bar{L}_n = L_{-n}$, and so one half of the conformal generators has been eliminated.

At the level of the stress tensor, the requirement is $T(\bar{z}) = \bar{T}(z)$. In Cartesian coordinates this reads $T_{xy} = 0$ on the real axis, the so-called **conformal boundary condition**. Its physical meaning is that there is no energy-momentum flow across $\mathbb{R}$.

This has important consequences on the conformal Ward identity where $T(z)$ is applied to a product of primary operators $X = \prod_j \phi_j(z_j, \bar{z}_j)$ situated in the upper half plane. The contour C surrounding all z_j can then be taken as a large semicircle with the diameter parallel to the real axis. However, writing the same identity for $\bar{T}(\bar{z})$ yields another Ward identity involving the conjugate semicircle contour $\bar{C}$, and since $\bar{T} = T$ when $z \in \mathbb{R}$, the two contours can be fused into a complete circle surrounding both z_j and $\bar{z}_j$. The end result is thus

$$T(z)X = \sum_j \left(\frac{\Delta_j}{(z - z_j)^2} + \frac{\partial_{z_j}}{z - z_j} + \frac{\bar{\Delta}_j}{(\bar{z} - \bar{z}_j)^2} + \frac{\partial_{\bar{z}_j}}{\bar{z} - \bar{z}_j} \right) X. \tag{4.52}$$

The fact that each of the usual terms occur twice means that everything happens as if each primary operator in the upper half plane were accompanied by a **mirror operator** in the lower half plane. This means that computations in the BCFT can be done using a **method of images** similar to that used in electrostatics when solving the Laplace equation with boundary conditions. Correlation functions are computed as if the theory were defined on the whole complex plane, and governed by a single Virasoro algebra: the physical operators are then situated in the upper half-plane, and their unphysical mirror images in the lower half-plane. The simplification of getting rid of $\bar{L}_n$ has thus been achieved at the price of doubling the number of points in correlation functions. In practice, the former simplification largely outweighs the latter complication.

In particular, the n-point boundary correlation functions satisfy the very same differential equations as $2n$-point bulk correlation functions, but with different boundary conditions. The most interesting cases are $n = 1$ and $n = 2$, both tractable in the bulk picture in several situations of practical importance. As examples of the physical information which can be extracted from these cases we should mention, for $n = 1$, the probability profile of finding a monomer of a loop at a certain distance from the boundary, and for $n = 2$, the probability that a polymer comes close to the boundary at two prescribed points. A particularly celebrated application of the $n = 2$ case is Cardy's computation of the *crossing probability* that a percolation cluster traverses a large rectangle, as a function of the aspect ration of the latter.

The radial quantisation scheme still makes sense in BCFT. The associated conformal mapping

$$w(z) = \frac{L}{\pi} \ln z \tag{4.53}$$

transforms the upper half plane into a semi-infinite strip of width L with non-periodic transverse boundary conditions. The two rims of the strip are then the images of the positive and the negative real axis, and the time (resp. space) direction is parallel (resp. perpendicular) to the axis of the strip. The dilatation operator reads $\mathcal{D} = L_0$ and the Hamiltonian $H = (\pi/L)(L_0 - c/24)$. Non-trivial eigenstates of H are formed by a **boundary operator** $\phi_j(0)$ acting on the vacuum state, $|\Delta\rangle = \phi_j(0)|0\rangle$.

In general, we expect boundary operators to have different scaling dimensions than bulk operators. This can be understood from the method of images: when a primary operator approaches the boundary it interacts with its mirror image and, by the usual OPE, produces a series of other primaries which then describe the boundary critical behaviour.

Likewise, a field $\phi_{(r,s)}$ with a given interpretation in the bulk will typically have a different interpretation when situated on the boundary. Examples pertinent to loop models will be given later.

Note that the usual transformation law

$$\phi'(w, \bar{w}) = \left(\frac{dw}{dz}\right)^{-\Delta} \left(\frac{d\bar{w}}{d\bar{z}}\right)^{-\bar{\Delta}} \phi(z, \bar{z}) \tag{4.54}$$

applied to a boundary operator is the reason why we have not discussed *finite* Dirichlet boundary conditions at the beginning of this section. More generally, any uniform boundary condition is expected to flow under the renormalisation group towards a *conformally invariant boundary condition*. It is one of the goals of BCFT to classify such boundary conditions.

One of the main results obtained is the following: For **diagonal models** (i.e., $n_{\Delta,\bar{\Delta}} = \delta_{\Delta,\bar{\Delta}}$) there is a bijection between the primary operator in the bulk CFT and the conformally invariant boundary conditions in the BCFT. For example, for the Ising model the three different bulk primary operators (the identity $I = \phi_{(1,1)}$, the spin $\sigma = \phi_{(1,2)}$, and the energy $\varepsilon = \phi_{(2,1)}$) correspond to three types of uniform boundary conditions in the lattice model of spins (fixed $s = +1$ and $s = -1$, and free boundary conditions).

To this point, we have discussed only uniform boundary conditions. It is important to realise that the radial quantisation picture with a boundary operator $\phi_j(0)$ situated at the origin is compatible also with mixed boundary conditions, i.e., one boundary condition on the negative real half-axis and another on the positive half-axis. In this case, $\phi_j(0)$ is called a **boundary condition changing operator**. One then needs a second operator $\phi_j(\infty)$ situated at infinity to change back the boundary condition. A more symmetric picture is obtained by mapping the upper half-plane to the strip, through (4.53). There are then different boundary conditions on the two sides of the strip, and a boundary condition changing operator is located at either end of the strip. More generally, one may study a BCFT on any simply connected domain with a variety of different boundary conditions along the boundary, each separated by a boundary condition changing operator.

For bulk CFT, crucial insight is gained by considering the theory on a torus. The analogous tool for BCFT is to consider the theory on an annulus.[6] In analogy with the torus case, we denote by $L \in \mathbb{R}$ the width of the annulus and by $M \in \mathbb{R}$ its length (in the periodic direction), defining $\tau = iM/L \in i\mathbb{R}$. The boundary conditions on the two rims are denoted, symbolically, a and b. Then

$$Z_{ab}(\tau) = \text{Tr}\left(q^{L_0 - c/24}\right) \tag{4.55}$$

with $q = \exp(\pi i \tau)$. (In the bulk case there is a further "barred" factor under the trace.) When expanded over the characters

$$Z_{ab}(\tau) = \sum_{\Delta} n_{\Delta}^{(ab)} \chi_{(c,\Delta)}(\tau), \tag{4.56}$$

one then obtains a linear (rather than quadratic) expression.

Equivalently, one might exchange the space and time direction and view the annulus as a cylinder of circumference ω_2 and finite length ω_1, with boundary conditions a (resp. b) in the initial (resp. final) state. This leads to

$$Z_{ab}(\tau) = \langle b | e^{\tau^{-1} \mathcal{H}_{\text{bulk}}} | a \rangle, \tag{4.57}$$

where now $\mathcal{H}_{\text{bulk}}$ is the Hamiltonian of the *bulk* CFT propagating between boundary states $|a\rangle$ and $\langle b|$. The links between bulk and boundary CFT result from a detailed study of the equivalence between (4.55) and (4.57).

---

[6]It makes sense to think of this in the radial quantisation, or transfer matrix, picture. The theories are initially considered on a semi-infinite cylinder (resp. a strip) with specified transverse boundary conditions (periodic, resp. non-periodic) and unspecified longitudinal boundary conditions. This gives access to the transfer matrix eigenvalues. To access the fine structure, such as amplitudes of the eigenvalues, one must impose periodic longitudinal boundary conditions and take the length of the cylinder (resp. strip) to be *finite*.

4.3.3 Coulomb Gas for Loops on an Annulus

Consider now a loop model defined on an annulus which we shall take as an $L \times M$ rectangle with coordinates $r_x \in [0, L]$ and $r_y \in [0, M]$. The boundary conditions are free (f) in the r_x-direction and periodic in the r_y-direction. Recently, Cardy has shown how to impose the correct marginality requirement for this geometry.

Consider first the continuum-limit partition function $Z = Z_{\mathrm{ff}}(\tau)$ from (4.55) in the limit $M/L \gg 1$ of a very long and narrow annulus. The modular parameters $\tau = iM/L$ and $q = \exp(i\pi\tau) = \exp(-\pi M/L)$. We expect in this limit that only the identity operator contributes to Z, and so

$$Z \sim q^{-c/24} \sim \exp\left(\frac{\pi c M}{24L}\right). \tag{4.58}$$

The central charge c is (4.37) from the bulk theory, and in particular is known to vary with the coupling constant g.

The question then arises how (4.58) is compatible with the continuum-limit action (4.29). According to Cardy, the answer is that there is a background magnetic flux m_0, a sort of electromagnetic dual of the background electric charge e_0 present in the cylinder geometry. Thus, in the continuum limit there is effectively a number (in general *fractional*) m_0 of oriented loops running along the rims of the annulus, giving rise to a height difference between the left and the right rim. Accepting this hypothesis, we can write

$$\phi(r_x, r_y) = \tilde{\phi}(r_x, r_y) + \frac{\pi m_0 r_x}{L} \tag{4.59}$$

where $\tilde{\phi}$ is a "gauged" height field that still contains the elastic fluctuations but obeys identical Dirichlet boundary conditions on both rims, say $\tilde{\phi}(0, r_y) = \tilde{\phi}(L, r_y) = 0$.

By the usual functional integrations, the free field $\tilde{\phi}$ contributes $\frac{q^{-1/24}}{P(q)}$ to Z, corresponding to $c = 1$. The last term in (4.59) modifies the action (4.29) by $\Delta S = \frac{g}{4\pi}(\pi m_0)^2 \frac{M}{L}$ and thus multiplies Z by a factor $e^{-\Delta S} = q^{g m_0^2/4}$, which correctly reproduces the contribution of the last term in (4.37) to (4.58) provided that we set

$$m_0 = \pm \frac{(1-g)}{g}. \tag{4.60}$$

This value of m_0 can be retrieved from a marginality requirement which has the double advantage of being more physically appealing and of not invoking the formula (4.37) for c. Indeed, if m_0 is too large a pair of oriented loop strands will shed from the rims, corresponding to a vortex pair of strength $m = \pm 2$ situated at the top and the bottom of the annulus. This vortex pair can then annihilate in order to reduce the free energy. And if m_0 is too small the opposite will occur. The equilibrium requirement is then that inserting such a vortex pair must be an exactly marginal perturbation in the RG sense, i.e., the corresponding *boundary* scaling dimension is $x_v = 1$.

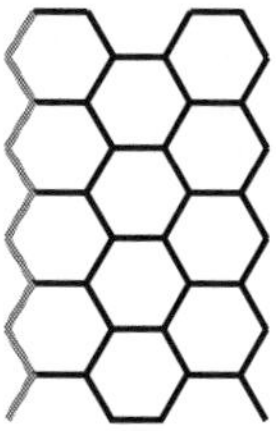

Fig. 4.8 Hexagonal lattice in
an annular geometry. The *top*
and the *bottom* of the figure
are identified. Boundary
edges on the *left* are shown in
grey [15]

The free energy increase for creating the vortex pair is, by the same gauge argument as before,

$$\Delta S = \frac{g}{4\pi} \left((m_0 \pm 2)^2 - m_0^2 \right) \left(\frac{\pi}{L} \right)^2 ML \tag{4.61}$$

and noting the factor of 24 between c and the scaling dimension x_v, we now have $e^{-\Delta S} = q^{-x_v}$ from (4.58), so that

$$x_v = \frac{g}{4} \left((m_0 \pm 2)^2 - m_0^2 \right) = 1 \tag{4.62}$$

and we recover (4.60). The ambiguity on the sign in (4.60) will be lifted in Sect. 4.3.5 below.

4.3.4 BCFT and the O(n) Model

The O(n) model with suitably modified surface couplings permits one to realise the ordinary, special, and extraordinary surface transitions described qualitatively in Sect. 4.3.1. To this end, one studies the model defined in the annular geometry of Sect. 4.3.3.

To be precise, the special transition requires the loops to be in the dilute phase, and so we shall assume this to be the case throughout Sect. 4.3.4. The results for the ordinary and extraordinary transitions hold true in the dense phase as well.

A well-studied case is the hexagonal-lattice loop model (4.18). The lattice is oriented such that one third of the lattice bonds are parallel to the x-axis, as shown in Fig. 4.8. The fugacity of a monomer is still denoted K in the bulk, but we now take a different weight K_s for a monomer touching the *left* rim of the annulus, $x = 0$. In contrast, the right rim of the annulus, $x = L$, enjoys free boundary conditions, meaning that its surface monomers still carry the usual weight K.

In this section we wish to limit the discussion to the case where only the left boundary sustains particular ($\neq$ free) boundary conditions; this is sometimes referred to as *mixed* boundary conditions. The case where both boundaries are distinguished is also of interest.

The loop model described above has been thoroughly studied by Batchelor and coworkers, in particular using Bethe Ansatz analysis. They find in particular that when $K_s = K$ the model is integrable and belongs to the universality class of the *ordinary* transition, while for

$$K_s = K_s^S = (2 - n)^{-1/4} \tag{4.63}$$

it is also integrable and describes the special transition.[7] This is consistent with a boundary RG scenario, where K_s^S is a repulsive fixed point that flows towards either of the attractive fixed points $K_s^O < K$ and $K_s^E = \infty$, the former (resp. latter) point describing the ordinary (resp. the extraordinary) transition.

This scenario is corroborated by a detailed analysis showing that a perturbation to the fixed point K_s^E is RG irrelevant. Moreover, the operator conjugate to K_s is obviously the energy density on the boundary. At the special transition, this operator can be identified with $\phi_{(1,3)}$ of weight $\Delta_{1,3} = \frac{2}{g} - 1$, and so this is a relevant perturbation (i.e., $\Delta_{1,3} < 1$) only for $g > 1$ (i.e., in the dilute phase). On the other hand, the surface energy density has weight $\Delta = 2$ at the ordinary transition, and so is always irrelevant.

4.3.5 Critical Exponents

Surface watermelon exponents can be defined as in Sect. 4.2.4, the only difference being that the ℓ legs are inserted at the boundary. We shall denote these exponents by x_ℓ^O, x_ℓ^S, x_ℓ^E at the ordinary, special, extraordinary surface transition respectively. Whenever a result applies to any of these transitions, we use the generic notation x_ℓ', where the prime indicates a surface rather than a bulk exponent.

For the ordinary transition, x_ℓ^O can be derived by a slight refinement of the marginality argument given in Sect. 4.3.3. First recall that in the continuum limit there is a background flux m_0 given by (4.60), corresponding to a (fractional) number of oriented loop strands running along the rims of the annulus. Suppose now that we wish to evaluate the scaling dimension x_ℓ^O corresponding to having $\ell > 0$ non-contractible oriented loop strands running around the periodic direction of the annulus. This can be done by evaluating the free energy increase $\Delta S = S_\ell - S_0$ due to these strands, as in (4.61)

$$\Delta S = \frac{g}{4\pi}\left((\ell + m_0)^2 - m_0^2\right)\left(\frac{\pi}{L}\right)^2 ML \tag{4.64}$$

and using $e^{-\Delta S} = q^{-x_\ell^O}$ from (4.58).

The question now arises which sign for m_0 to pick in (4.60). With the plus sign we would have $x_2 = 1$ independently of g, in clear contradiction with numerical results. Taking therefore the minus sign leads to the result

$$x_\ell^O = \frac{1}{4}g\ell^2 - \frac{1}{2}(1 - g)\ell. \tag{4.65}$$

The derivation just presented follows the argument of Cardy, but in fact (4.65) was found a long time before by other means. Duplantier and Saleur were the first to

---

[7]Technically speaking this is the mixed ordinary-special transition, but we have simplified the terminology according to the above remarks.

propose (4.65) for any ℓ, by noting that their numerical transfer matrix results were in excellent agreement with the following locations in the Kac table

$$x_\ell^O = \begin{cases} \Delta_{1,1+\ell} & \text{for the dense O}(n) \text{ model} \\ \Delta_{1+\ell,1} & \text{for the dilute O}(n) \text{ model} \end{cases} \quad (4.66)$$

from which (4.65) follows by the identification (4.40). On a more rigorous level, (4.65) has been established by Bethe Ansatz (BA) techniques.

For the special transition, x_ℓ^S does not seem to permit a CG derivation. It is however known from the BA analysis that one has

$$x_\ell^S = \frac{1}{4} g(1+\ell)^2 - (1+\ell) + \frac{4 - (1-g)^2}{4g}$$
$$= \Delta_{1+\ell,2} \quad \text{for the dilute O}(n) \text{ model} \quad (4.67)$$

in this case.

Alternatively, one may imagine producing the special ℓ-leg operator $\mathcal{O}_\ell^S$ by fusion of the ordinary ℓ-leg operator $\mathcal{O}_\ell^O$ and an ordinary-to-special boundary condition changing operator ϕ_{OS}. The scaling dimension (4.67) pertains to the insertion of this composite operator at either strip end. Comparing the Kac indices in (4.66) and (4.67), and using the CFT fusion rules, immediately leads to the identification $\phi_{OS} = \phi_{1,2}$. If one wants special boundary conditions on both the left and the right rim, two insertions of ϕ_{OS} are needed (to change from special to ordinary and back again). One would then expect $\Delta_{1+\ell,3}$, as is indeed confirmed by the BA analysis.

Finally, the extraordinary transition is rather trivially related to the ordinary transition. Indeed, for $K_s = \infty$ the entire left rim of the annulus will be coated by a straight polymer strand, so that the remaining system (of width $L - 1$) effectively sees free boundary conditions—this is sometimes called the **teflon effect**. Thus, for $\ell = 0$ the coating strand will be the left half of a long stretched-out loop, whose right half will act as a one-leg operator, and one effectively observes the exponent x_1^O. For $\ell > 0$, one of the legs will act as the coating strand, and one observes $x_{\ell-1}^O$.

4.4 Temperley-Lieb Algebra

Consider the loop model corresponding to the $Q = n^2$ state Potts model on the square lattice. The loops live on a tilted square lattice, which we think of as being built up by a row-to-row transfer matrix.

We impose periodic boundary conditions in the time direction, so that the topology is that of an annulus. The loops can then have two different homotopies with respect to the periodic direction: contractible or non-contractible. In general, we may define a statistical ensemble by giving a weight n to each contractible loop and a weight ℓ to each non-contractible loop. The weight of the configuration in Fig. 4.9 is then $n\ell^2$. The model just defined has the algebraic structure of the Temperley-Lieb algebra.

A more general model in which special weights n_1 and ℓ_1 are given to loops that touch at least once the left boundary has been studied by Jacobsen and Saleur. The

Fig. 4.9 Configuration of
self-avoiding fully-packed
loops on an annulus [18]

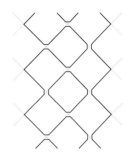

algebraic structure of that model is that of the so-called one-boundary Temperley-Lieb (1BTL) algebra. One can similarly define a 2BTL loop model with 8 different loop weights, depending on their homotopy and which boundaries they touch. This has been studied in details by Dubail, Jacobsen, and Saleur.

4.4.1 Usual TL Algebra

We now consider in details the case of the usual TL algebra; we shall sometimes find it convenient to refer to it as the 0BTL algebra.

Consider a system of N strands labelled $i = 1, 2, \ldots, N$. The lattice is built up from elementary generators e_i, acting on strands i and $i + 1$, as shown in Fig. 4.10. More precisely, in the selfdual (hence critical, if $0 \leq n \leq 2$) case where all *local* vertex weights are unity, the transfer matrix reads

$$T = \left(\prod_{j=1}^{N/2-1} (1 + e_{2j}) \right) \left(\prod_{j=1}^{N/2} (1 + e_{2j-1}) \right). \tag{4.68}$$

The generators e_i satisfy the relations

$$e_i^2 = n e_i$$
$$e_i e_{i\pm1} e_i = e_i \tag{4.69}$$
$$[e_i, e_j] = 0 \quad \text{for } |i - j| \geq 2$$

which can be verified simply by drawing what they mean in terms of loops.

The identity and the $N - 1$ generators e_i define the **Temperley-Lieb algebra** $TL_N(n)$, subject to the above relations. Graphically, the application of the last two relations allows to deform and diminish the size of a loop, and when it has reached its minimal possible size it can be taken away and replaced by the weight n due to the first relation.

4.4.1.1 States and Transfer Matrix Decomposition

The transfer matrix T acts on states which can be depicted graphically as non-crossing link patterns within a slab bordered by two horizontal rows, each of N

Fig. 4.10 From *left* to *right*: identity I and Temperley-Lieb generator e_i acting on two strands i and $i + 1$; *left* and *right* boundary identity operator [18]

points. The complete list of states for $N = 4$ is shown in Fig. 4.11. The bottom (resp. top) row of the slab corresponds to time $t = 0$ (resp. $t = t_0$); the transfer matrix propagates the states from t_0 to $t_0 + 1$ and thus acts on the top of the slab only.

A link joining the top and the bottom of the slab is called a **string**, and any other link is called an *arc*. We denote by s the number of strings in a given state. Any state can be turned into a pair of **reduced states** by cutting all its strings and pulling apart the upper and lower parts. For convenience, a cut string will still be called a string with respect to the reduced state. The complete list of reduced states for $N = 4$ is shown in Fig. 4.12.

Conversely, a state can be obtained by adjoining two reduced states, gluing together their strings in a unique fashion. Thus, if we define d_{2j} as the number of reduced states with $s = 2j$ strings, the number of states with $s = 2j$ strings is simply d_{2j}^2.

The partition function $Z_{N,M}$ on an annulus of width N strands and height M units of time cannot be immediately expressed in terms of reduced states only, since these do not contain the information about how many loops (contractible or noncontractible) are formed when the periodic boundary condition is imposed. We can however write it in terms of states as

$$Z_{N,M} = \langle u | T^M | v \rangle. \tag{4.70}$$

At time $t_0 = 0$ the top and the bottom of the slab must be identified. Therefore, the entries of the right vector $|v\rangle$ are one whenever the corresponding state contains no arcs, and each of its links connects a point in the bottom row to the point immediately above it in the top row; all other entries of $|v\rangle$ are zero. At time $t_0 = M$ the top and the bottom of the slab must be re-glued. Therefore, the left vector $\langle u |$ is obtained by identifying the top and bottom rows for each state; counting the number of loops of each type gives the corresponding weight as a monomial in the loop weights n and ℓ.

The reduced states can be ordered according to a decreasing number of strings. The states can be ordered first according to a decreasing number of strings, and next, for a fixed number of strings, according to its bottom half reduced state. These orderings are brought out by the rows in Figs. 4.11–4.12.

With this ordering of the states, T has a block-wise lower triangular structure in the basis of reduced states, since the generator e_i can annihilate two strings (if their position on the top of the slab are i and $i + 1$) but cannot create any strings.

In the basis of states, T is block-wise lower triangular with respect to the number of strings, for the same reason. Each block on the diagonal in this decomposition corresponds to a definite number of strings. The block corresponding to $s = 2j$ strings is denoted $\tilde{T}_j$. But since T acts only on the top of the slab, each $\tilde{T}_j = T_j \oplus$

Fig. 4.11 List of all 0BTL states on $N = 4$ strands. *Each row* corresponds to a definite sector of the transfer matrix [18]

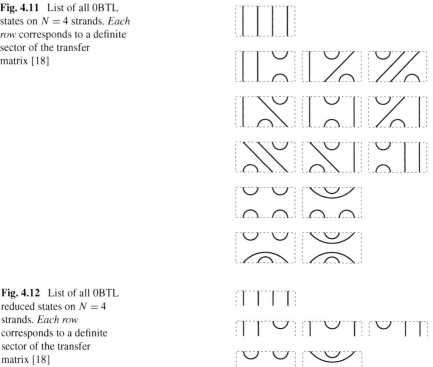

Fig. 4.12 List of all 0BTL reduced states on $N = 4$ strands. *Each row* corresponds to a definite sector of the transfer matrix [18]

$\dots \oplus T_j$ is in turn a direct sum of d_j identical blocks T_j which correspond simply to the action of T on the reduced states with $2j$ strings.

In particular, the eigenvalues of T are the union of the eigenvalues of T_j, where the T_j now act in the much smaller basis of reduced states. This observation is particularly useful in numerical studies.

4.4.1.2 The Dimensions d_L and D_L

In spite of the periodic boundary conditions, $Z_{N,M}$ is obviously not a usual matrix trace. It can however be decomposed on standard traces by a combinatorial construction due to Richard and Jacobsen. The generalisation to the case with boundaries was obtained by Jacobsen and Saleur.

We take for now the width of the annulus $N = 2N_2$ to be even. For each transfer matrix block T_j we define the corresponding character as

$$K_j = \text{Tr}(T_j)^M, \tag{4.71}$$

where we stress that the trace is over *reduced* states. Obviously we have

$$K_j = \sum_{i=1}^{d_{2j}} (\lambda_i^{(j)})^M, \tag{4.72}$$

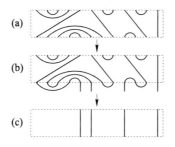

Fig. 4.13 Construction of invariant reduced states. (**a**) A configuration contributing to Z_2 with $N_2 = 6$, here depicted as a state. (**b**) Application on the *bottom* of the reduced state corresponding to the *top half* of (**a**). (**c**) After removal of the arcs one has simply $2j$ links

where $\lambda_i^{(j)}$ are the eigenvalues of T_j. The expression of the partition function in terms of transfer matrix eigenvalues is more involved, due essentially to the non-local nature of the loops, and reads

$$Z_{N,M} := \sum_{j=0}^{N_2} Z_j = \sum_{j=0}^{N_2} D_{2j} K_j, \tag{4.73}$$

where Z_j is the annulus partition function constrained to have exactly $L = 2j$ non-contractible loops, and D_{2j} are some eigenvalue amplitudes to be determined. To be more precise, we decompose Z_j in terms of K_k as follows

$$Z_j = \sum_{k=j}^{N_2} D(k, j)\ell^{2k} K_k$$

$$D_{2j} = \sum_{i=0}^{j} D(j, i)\ell^{2i} \tag{4.74}$$

and consider next the inverse decomposition

$$K_k = \sum_{j=k}^{N_2} E(j, k)\frac{Z_j}{\ell^{2j}}. \tag{4.75}$$

The determination of the coefficients $E(j, k)$ can be turned into a combinatorial counting problem as follows. First, recall that the characters K_k were defined as *traces* over reduced states. We must now determine how many times each Z_j occurs within a given trace. Consider therefore some configuration $\mathscr{C}$ on the annulus that contributes to Z_j. An example with $j = 2$ and $N_2 = 6$ is shown in Fig. 4.13a. It is convenient not to represent the contractible loops within the configuration, i.e., to depict it as a state. This configuration will contribute to the trace only over such reduced states $\mathscr{S}$ that are left *invariant* by the action of the configuration. Therefore, $\mathscr{S}$ must contain the same arcs as does $\mathscr{C}$ in its top row (see Fig. 4.13b). It suffices therefore to determine the parts of $\mathscr{S}$ which connect onto the starting points of the $2j$ non-contractible lines (see Fig. 4.13c). Since the goal is to determine the contribution to K_k, precisely $2k$ strings and $j - k$ arcs must be used. In other words, $E(j, k)$ is precisely the number of reduced states on $2j$ strands, and using $2k$ strings.

Now let

$$E^{(k)}(z) = \sum_{j=0}^{\infty} E(j, k) z^j \tag{4.76}$$

be the corresponding generating function, where z is a formal parameter representing the weight of an arc, or of a pair of strings. When $k = 0$, a reduced state with no strings is either empty, or has a leftmost arc which divides the space into two parts (inside the arc and to its right) each of which can accommodate an independent arc state. The generating function $f(z) = E^{(0)}(z)$ therefore satisfies $f(z) = 1 + zf(z)^2$ with regular solution

$$f(z) = \frac{1 - \sqrt{1 - 4z}}{2z} = \sum_{j=0}^{\infty} \frac{(2j)!}{j!(j+1)!} z^j. \tag{4.77}$$

When $k \neq 0$, the strings simply divide the space into $2k + 1$ parts each of which contains an independent arc state. Therefore,

$$E^{(k)}(z) = z^k f(z)^{2k+1} = \sum_{j=k}^{\infty} \left[\binom{2j}{j+k} - \binom{2j}{j+1+k} \right] z^j \tag{4.78}$$

and in particular we have

$$d_L = E\left(\frac{N}{2}, \frac{L}{2}\right) = \binom{N}{(N+L)/2} - \binom{N}{1 + (N+L)/2}. \tag{4.79}$$

Note that d_L depends on N, but we usually will not mention this explicitly.

Inversion of the linear system (4.75) finally leads to

$$D(j, k) = (-1)^{j+k} \binom{j+k}{2k}, \tag{4.80}$$

which can also be written

$$D_L = U_L(\ell/2), \tag{4.81}$$

where $U_k(x)$ is the well-known k^{th}-order Chebyshev polynomial of the second kind, $U_L(\cos\theta) = \frac{\sin(L+1)\theta}{\sin\theta}$.

The total number of states is

$$\sum_{j=0}^{N_2} d_{2j} = \binom{N}{N/2}. \tag{4.82}$$

One should also note the sum rule

$$\sum_{j=0}^{N_2} d_{2j} D_{2j} = \ell^N \tag{4.83}$$

which expresses the fact that there are ℓ degrees of freedom living on each site.

The representation-theory of the TL algebra is well-known. For generic values of n, the irreducible representations are labelled by a single integer $L = 0, 2, \ldots, N$ which counts the number of non-contractible (or "through") lines, and have dimension equal to the multiplicity of the spin $\frac{L}{2}$ representation in a chain of N spins $1/2$. This dimension is easily seen to be d_L of (4.79). Meanwhile, D_L is a q-dimension for the corresponding commutant, which is the **quantum algebra** $U_q(sl_2)$ with $q + q^{-1} = \ell$.

4.4.2 One-Boundary TL Algebra

In the one-boundary case, contractible loops touching at least once the left boundary receive a weight n_1 which is different from the weight n of a bulk loop. Coding this algebraically requires the introduction of an additional generator b_1 acting on the (left) boundary, such that

$$b_1^2 = b_1$$
$$e_1 b_1 e_1 = n_1 e_1 \tag{4.84}$$
$$[b_1, e_i] = 0 \quad \text{for } i = 2, 3, \ldots, N-1.$$

These relations, with (4.69), define the **one-boundary Temperley-Lieb** (1BTL) **algebra**.

Graphically, the action of b_1 can be depicted by adding a **blob** (shown in the following figures as a circle) to the link that touches the boundary. The first relation in (4.84) means that all the anchoring points of a boundary touching loop, except the last one, can be taken away. The third relation and (4.69) allow to deform and diminish the size of a boundary loop (while keeping it glued to one of its anchoring points on the boundary), and when it has reached its minimal possible size it can be taken away and replaced by the weight n_1 due to the second relation of (4.84).

The transfer matrix can be taken as

$$T = \left(\prod_{j=1}^{N/2-1} (1 + e_{2j}) \right) \left(\prod_{j=1}^{N/2} (1 + e_{2j-1}) \right) (\lambda_1 1 + b_1) \tag{4.85}$$

where a non-zero value of λ_1 would mean that with some probability a loop may come close to the boundary without actually touching it. We shall mostly set $\lambda_1 = 0$ in what follows. This has the advantage of reducing the dimension of the space on which T acts, since then the leftmost link in any (reduced) state may be taken to be blobbed. The algebraic results for the case $\lambda_1 \neq 0$ are simply related to those for the case $\lambda_1 = 0$, and we shall discuss them in due course.

4.4.2.1 States and Transfer Matrix Decomposition

The states of the transfer matrix are as in the 0BTL case, except that links which are exposed to the boundary (i.e., which are not to the right of the leftmost string) may

Fig. 4.14 List of all 1BTL
states on $N = 4$ strands (with
$\lambda_1 = 0$). *Each row*
corresponds to a definite
sector of the transfer
matrix [18]

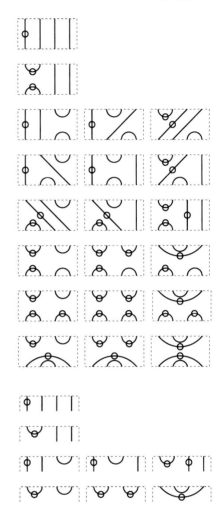

Fig. 4.15 List of all 1BTL
reduced states on $N = 4$
strands (with $\lambda_1 = 0$). *Each
row* corresponds to a definite
sector of the transfer
matrix [18]

be blobbed. Also, any link touching the leftmost site ($i = 1$) is necessarily blobbed,
since we have taken $\lambda_1 = 0$ in (4.85).

The states for $N = 4$ are shown in Fig. 4.14, and the corresponding reduced
states are given in Fig. 4.15. When a blobbed and an un-blobbed link are adjoined
(e.g., when transferring, or when forming an inner product) the result is a blobbed
(restricted) link.

The decomposition of the transfer matrix into sectors (blocks) takes place exactly
as in the 0BTL case, with one important addition. Namely, once the number of
strings $s = 2j$ has been fixed, the blocks T_j are block-wise 2×2 lower triangular
with respect to the blobbing status of the leftmost string. Indeed, acting by b_1 can
blob the leftmost string, but a string—*qua* a conserved object with respect to T_j—
cannot subsequently be un-blobbed. The elementary blocks are therefore T_j^{b} and
T_j^{u}, where the superscript indicates the blobbing status (b for blobbed, and u for

un-blobbed) of the leftmost string. With $\lambda_1 = 0$, there is no un-blobbed sector with $j = N/2$, and by convention the sector with $j = 0$ is un-blobbed (for any λ_1). In Figs. 4.14–4.15, the second rows give the unique un-blobbed state with $j = 1$.

4.4.2.2 The Dimensions d_L^α and D_L^α

Using similar combinatorial techniques as in the usual TL case, one establishes that

$$d_L^{\mathrm{u}} = d_L^{\mathrm{b}} = \binom{N}{(N-L)/2} \quad \text{(for } \lambda_1 \neq 0) \tag{4.86}$$

so that the total number of states is

$$\sum_{j=0}^{N_2} d_{2j}^{\mathrm{u}} + \sum_{j=1}^{N_2} d_{2j}^{\mathrm{b}} = 2^N. \tag{4.87}$$

After some work on generating function one arrives at

$$\begin{aligned} D_L^{\mathrm{u}} &= U_L(\ell/2) - \ell_1 U_{L-1}(\ell/2) \\ D_L^{\mathrm{b}} &= \ell_1 U_{L-1}(\ell/2) - U_{L-2}(\ell/2) \end{aligned} \tag{4.88}$$

and the sum-rule becomes

$$\sum_{j=0}^{N_2} d_{2j}^{\mathrm{u}} D_{2j}^{\mathrm{u}} + \sum_{j=1}^{N_2} d_{2j}^{\mathrm{b}} D_{2j}^{\mathrm{b}} = \ell^N. \tag{4.89}$$

4.5 Exact CFT Partition Functions

By combining the main results of the preceding two sections, it is possible to construct the exact continuum-limit partition functions of the loop models defined on an annulus of size $L \times M$. The periodic direction is that of size M.

4.5.1 0BTL Loop Model

We recall that this model is defined by giving a weight $n = -2\cos(\pi g)$ to each contractible loop, and a (possibly different) weight $\bar{n} = 2\cos(\pi e_0)$ to each non-contractible loop.

According to (4.56) we have

$$Z := Z_{\mathrm{ff}}(q) = \sum_\Delta n_\Delta \chi_{(c,\Delta)}(q) \tag{4.90}$$

where the sum is over the boundary scaling dimensions $x = \Delta$. Here $\chi_{(c,h)}(q)$ is the generic character

$$\chi_{(c,\Delta)}(\tau) = \mathrm{Tr}\, q^{L_0 - c/24} = \frac{q^{\Delta - c/24}}{P(q)}, \tag{4.91}$$

where the latter expression stems from the fact that the number of descendents of $|\Delta\rangle$ at level k in a generic Virasoro module is just the number of integer partitions of k. Explicitly

$$\frac{1}{P(q)} := \prod_{k=1}^{\infty} \frac{1}{1 - q^k} = \sum_{k=0}^{\infty} p(k) q^k. \tag{4.92}$$

We recall that the modular parameter is $q = \exp(i\pi\tau) = \exp(-\pi M/L)$. The degeneracy factor n_Δ states how many times a given character appears in the partition function, and as usual for non-minimal theories it needs not in general be an integer. We omit in the following the subscript ff which reminds us that the boundary conditions on both rims of the annulus are free.

The CFT partition function is then

$$Z[g, e_0] = \frac{q^{-c/24}}{P(q)} \sum_{\ell \in \mathbb{Z}} \frac{\sin((1 + \ell)\pi e_0)}{\sin(\pi e_0)} q^{\frac{g\ell^2}{4} - \frac{(1-g)\ell}{2}}. \tag{4.93}$$

The expression (4.93) was first obtained by Saleur and Bauer, using techniques of integrability and quantum groups. It was later re-derived and discussed by Cardy from a Coulomb gas point of view. We hold by now all the necessary ingredients to prove this relation:

1. The front factor $\frac{q^{-c/24}}{P(q)}$ is the usual contribution from the free boson, viz., the "gauged" height field $\tilde{\phi}$ of (4.59).
2. The $q^{x_\ell^o}$ factor codes the critical exponents of the ℓ-leg (watermelon) operators at the ordinary surface transition.
3. The degeneracy factor

$$n_\ell = U_\ell(\bar{n}/2) = \frac{\sin((1 + \ell)\pi e_0)}{\sin(\pi e_0)} \tag{4.94}$$

comes from the algebraic decomposition of the TL (Markov) trace over ordinary matrix traces.
4. The sum $\sum_{\ell \in \mathbb{Z}}$ is over the number of non-contractible lines on the annulus.

The attentive reader may object that (1) the expansion (4.90) should not be over generic characters, but the degenerate ones

$$K_{r,s} = \frac{q^{\Delta_{r,s}} - q^{\Delta_{r,-s}}}{q^{c/24} P(q)}, \tag{4.95}$$

and (2) the sum in (4.93) should be over $\ell \geq 0$ and not $\ell \in \mathbb{Z}$. While these observations are certainly correct, a little analysis shows that taking into account (1) and (2) leads to exactly the same result (4.93).

4.5.2 A Percolation Crossing Formula

The result (4.93) contains a lot of precious information in a very compact form. To illustrate the scope of this expression, we consider the limit $n \to 1$, which corresponds to bond percolation on the square lattice. In this case $c = 0$.

The partition function itself is $Z[g = \frac{2}{3}, e_0 = \frac{1}{3}]$. The part of (4.93) under the summation is

$$\sum_{\ell \in \mathbb{Z}} \frac{\sin((1+\ell)\pi/3)}{\sin(\pi/3)} q^{\frac{\ell^2}{6} - \frac{\ell}{6}}. \tag{4.96}$$

The contributions are only non-zero in the following cases

$$\begin{aligned}
\ell &= 6r: & q^{6r^2 - r} \\
\ell &= 6r - 2: & -q^{6r^2 - 5r + 1} \\
\ell &= 6r + 1: & q^{6r^2 + r} \\
\ell &= 6r + 3: & -q^{6r^2 + 5r + 1}.
\end{aligned}$$

Let us recall the **Euler pentagonal number theorem**:

$$P(q) = \prod_{k=1}^{\infty} (1 - q^k) = \sum_{k=-\infty}^{\infty} (-1)^k q^{k(3k-1)/2}. \tag{4.97}$$

A term with even $k = 2r$ reads $q^{6r^2 - r}$, and a term with odd $k = 2r + 1$ reads $q^{6r^2 + 5r + 1}$. Thus regrouping the contributions with $\ell = 6r$ and $\ell = 6r + 3$ the above sum is simply $P(q)$. One finds the same result by regrouping the contributions with $\ell = 6r - 2$ and $\ell = 6r + 1$.

So seemingly $Z[g = \frac{2}{3}, e_0 = \frac{1}{3}] = 2$. But taking into account that the equivalence between the TL loop model and the Potts model requires an even number of strands N—whence also ℓ is even—we have simply

$$Z\left[g = \frac{2}{3}, e_0 = \frac{1}{3}\right] = 1. \tag{4.98}$$

Consider now the probability p that a percolation cluster connects the two rims of the annulus. This happens if and only if there are no loops wrapping around the periodic direction. Such loops can be suppressed by setting $e_0 = \frac{1}{2}$. In view of the trivial normalisation ($Z = 1$) we have then

$$\begin{aligned}
p &= Z\left[g = \frac{2}{3}, e_0 = \frac{1}{2}\right] \\
&= \frac{1}{P(q)} \sum_{\ell \in \mathbb{Z}} \sin\left((1+\ell)\frac{\pi}{2}\right) q^{\frac{p(p-1)}{6}}.
\end{aligned}$$

The degeneracy factor is $+1$ if $\ell = 4r$ and -1 if $\ell = 4r + 2$. Thus

$$p = \frac{1}{P(q)} \sum_{r \in \mathbb{Z}} (q^{\frac{4r(4r-1)}{6}} - q^{\frac{(4r+2)(4r+1)}{6}}).$$

This can in turn be rewritten by using the **Jacobi triple product formula**

$$\sum_{k \in \mathbb{Z}} (-1)^k a^k q^{\frac{k(k-1)}{2}} = \prod_{k=1}^{\infty} \left(1 - aq^{k-1}\right)\left(1 - a^{-1}q^n\right)\left(1 - q^n\right) \tag{4.99}$$

in terms of the **Dedekind function** $\eta(\tau) = q^{1/24}(P(q))^{-1}$ as

$$p = \frac{\eta\left(-\frac{1}{3\tau}\right)\eta\left(-\frac{4}{3\tau}\right)}{\eta\left(-\frac{1}{\tau}\right)\eta\left(-\frac{2}{3\tau}\right)} = \sqrt{\frac{3}{2}} \frac{\eta(3\tau)\eta\left(\frac{3\tau}{4}\right)}{\eta(\tau)\eta\left(\frac{3\tau}{2}\right)}. \tag{4.100}$$

For a thin annulus, $q = \exp(-\pi M/L) \to 0$, we have $1 - p \sim q^{1/3}$. In terms of the conjugate modulus, $\tilde{q} = \exp(-2\pi L/M)$, a long cylinder corresponds to $\tilde{q} \to 0$. In that limit

$$p \sim \sqrt{\frac{3}{2}} \tilde{q}^{\frac{5}{48}}, \tag{4.101}$$

where we recognise the magnetic exponent of the $Q \to 1$ state Potts model. The result (4.100) can be seen as expressing all corrections to scaling for this well-known result.

4.5.3 1BTL Loop Model

Still working on the annulus, Jacobsen and Saleur have defined a more general model in which bulk loops have fugacity n or $\bar{n}$, and loops touching the left boundary have weight n_1 or $\overline{n_1}$, where in all cases the overline refers to non-contractible loops (i.e., loops that are not homotopic to a point). This is illustrated in Fig. 4.16.

We have seen above how the transfer matrix T of any loop model on the annulus can be decomposed into blocks T_ℓ labelled by the number of non-contractible loops ℓ. For the 1BTL loop model one may further decompose T_ℓ into the blobbed (resp. un-blobbed) sector T_ℓ^b (resp. T_ℓ^u) in which the leftmost non-contractible loop is required (resp. forbidden) to touch the left rim of the annulus. Indeed, since a non-contractible loop is conserved by definition, once it has been blobbed (i.e., touched the boundary) it cannot subsequently be un-blobbed. Therefore, T_ℓ is upper block-triangular in the basis $\{|b\rangle, |u\rangle\}$ and the previous argument applies *mutatis mutandis*.

Parameterising

$$n = 2\cos\gamma$$
$$n_1 = \frac{\sin[(r_1 + 1)\gamma]}{\sin(r_1\gamma)} \tag{4.102}$$

the central charge is (4.38) with $\gamma = \frac{\pi}{m+1}$. The parameter $r_1 \in (0, m+1)$ is in general a real number. The watermelon exponents are then

$$x_\ell^0(n, n_1) = \Delta_{r_1, r_1 \pm \ell} \tag{4.103}$$

where the upper (resp. lower) sign is for the blobbed (resp. un-blobbed) sector.

Fig. 4.16 Continuum-limit view of the four different types of loops, distinguished by their *colours*, in the 1BTL loop model. In this figure the annulus has been conformally mapped to the plane, and the "*left rim*" referred to in the text has become the *outer rim*. Contractible (resp. non-contractible) loops are those that do not (resp. that do) wrap around the *hole in the annulus*. Bulk (resp. *boundary*) loops are those that do not (resp. that do) touch the *outer rim* at least once. *Each bulk* (resp. *boundary*) loop has weight n (resp. n_1) if it is contractible, and $\bar{n}$ (resp. $\overline{n_1}$) if it is non-contractible

The result (4.103) follows integrability results for the corresponding XXZ spin chain with boundary terms when r_1 and m are integers. There is ample numerical evidence that it holds also for non-integer values. Deriving (4.103) from Coulomb gas arguments is an interesting open problem.

The 1BTL loop model contains the ordinary O(n) loop model as the special case $n_1 = n$, but it is clear that its transfer matrix must contain many more states in order to produce the correct weights for $n_1 \neq n$. Therefore, the conformal towers must be more densely filled, and the spectrum generating functions must contain fewer degeneracies. Since the loop model characters (4.95) contain just one subtraction, it seems reasonable that the 1BTL characters for generic $n_1 \neq n$ will not involve any subtractions, i.e., they must be the generic characters (4.91). This is indeed confirmed by numerical diagonalisation of the transfer matrix. Combining this with the result for the conformal weights (4.103), we conclude that the spectrum-generating functions for the blobbed and un-blobbed sectors read

$$Z_\ell^b = \frac{q^{\Delta_{r,r+\ell}-c/24}}{P(q)}, \qquad Z_\ell^u = \frac{q^{\Delta_{r,r-\ell}-c/24}}{P(q)}. \qquad (4.104)$$

To find out how to combine these sectors to obtain the complete partition function Z, one simply uses the multiplicities D^b and D^u derived in the preceding section. Parameterising the weights of non-contractible loops as

$$\bar{n} = 2\cosh\alpha, \qquad \overline{n_1} = \frac{\sinh(\alpha+\beta)}{\sinh\beta} \qquad (4.105)$$

the result reads

$$D_\ell^b = \frac{\sinh(\ell\alpha+\beta)}{\sinh\beta}, \qquad D_\ell^u = \frac{\sinh(\ell\alpha-\beta)}{\sinh(-\beta)}. \qquad (4.106)$$

Supposing L is even, and setting $\ell = 2j$, the results (4.104) and (4.106) lead to

$$Z = q^{-c/24}\left[\sum_{j=0}^{\infty}\frac{\sinh(2j\alpha + \beta)}{\sinh\beta}\frac{q^{\Delta_{r,r+2j}}}{P(q)} - \sum_{j=1}^{\infty}\frac{\sinh(2j\alpha - \beta)}{\sinh\beta}\frac{q^{\Delta_{r,r-2j}}}{P(q)}\right].$$

(4.107)

4.5.4 2BTL Loop Model

The 1BTL loop model can be generalised to the case where both boundaries of the annulus are distinguished. In this 2BTL loop model, bulk loops have a weight n, while boundary loops touching only the left (resp. right) boundary have weight n_1 (resp. n_2), and loops touching both boundaries have weight n_{12}.

This model is equivalent to a Potts model in which bulk spins have $Q = n^2$ states, while spins on the left (resp. right) boundary are constrained to a smaller number $Q_1 = nn_1$ (resp. $Q_2 = nn_2$) of states, of which there are $Q_{12} = nn_{12}$ common states.

The following parameterisation turns out to be instrumental for further study:

$$\begin{aligned}
n &= 2\cos\gamma\\
n_1 &= \frac{\sin[(r_1 + 1)\gamma]}{\sin(r_1\gamma)}\\
n_2 &= \frac{\sin[(r_2 + 1)\gamma]}{\sin(r_2\gamma)}\\
n_{12} &= \frac{\sin[(r_1 + r_2 + 1 - r_{12})\frac{\gamma}{2}]\sin[(r_1 + r_2 + 1 - r_{12})\frac{\gamma}{2}]}{\sin(r_1\gamma)\sin(r_2\gamma)}.
\end{aligned}$$

(4.108)

The full meaning of this parameterisation only becomes clear within the representation theory of the underlying algebra.

When defining the 2BTL loop model on an annulus of *even* width L, a non-contractible loop cannot touch both rims of the annulus. We thus need only the following additional three weights for non-contractible loops:

$$\begin{aligned}
\bar{n} &= 2\cos\chi\\
\bar{n}_1 &= \frac{\sin[(u_1 + 1)\chi]}{\sin(u_1\chi)}\\
\bar{n}_2 &= \frac{\sin[(u_2 + 1)\chi]}{\sin(u_2\chi)}.
\end{aligned}$$

(4.109)

The exact continuum limit partition function, expressed in terms of all these seven weights, has been derived by Dubail, Jacobsen and Saleur:

$$Z = \frac{q^{-c/24}}{P(q)} \sum_{n \in \mathbb{Z}} q^{\Delta_{r_{12}-2n, r_{12}}}$$

$$+ \frac{q^{-c/24}}{P(q)} \sum_{j \geq 1} \sum_{n \geq 0} \frac{\sin[(u_1 + u_2 - 1 + 2j)\chi] \sin \chi}{\sin(u_1\chi) \sin(u_2\chi)} q^{\Delta_{r_1+r_2-1-2n, r_1+r_2-1+2j}}$$

$$+ \frac{q^{-c/24}}{P(q)} \sum_{j \geq 1} \sum_{n \geq 0} \frac{\sin[(-u_1 + u_2 - 1 + 2j)\chi] \sin \chi}{\sin(-u_1\chi) \sin(u_2\chi)} q^{\Delta_{-r_1+r_2-1-2n, -r_1+r_2-1+2j}}$$

$$+ \frac{q^{-c/24}}{P(q)} \sum_{j \geq 1} \sum_{n \geq 0} \frac{\sin[(u_1 - u_2 - 1 + 2j)\chi] \sin \chi}{\sin(u_1\chi) \sin(-u_2\chi)} q^{\Delta_{r_1-r_2-1-2n, r_1-r_2-1+2j}}$$

$$+ \frac{q^{-c/24}}{P(q)} \sum_{j \geq 1} \sum_{n \geq 0} \frac{\sin[(-u_1 - u_2 - 1 + 2j)\chi] \sin \chi}{\sin(-u_1\chi) \sin(-u_2\chi)} q^{\Delta_{-r_1-r_2-1-2n, -r_1-r_2-1+2j}}.$$

(4.110)

The five-term structure of this expression permits one to read off the principal critical exponents. The trigonometric factors inside the four last terms are the eigenvalue amplitudes, which can be derived by generalising the combinatorial derivation of the preceding section.

Obviously, an expression like (4.110) contains a wealth of exact probabilistic information, which can be extracted explicitly for any special case of interest (such as percolation). Moreover, it determines the complete operator content of the two-boundary model, and the precise fusion rules of two one-boundary CBL type boundary condition changing operators.

4.6 Notes and References

General background for the topics of this chapter is given in [15, 27] and the 'yellow bible' [10].

The Q-states Potts model [29] and its different re-formulations for continuous values of Q described in Sect. 1 appeared in [2, 11, 14, 20], and of course in [1]. The Coulomb gas construction described in Sect. 2 arose in [19] and was worked out in [7, 8, 24, 25], see also the review [26]. The relationship with $2D$ CFT was made more precise in [9] and applications to polymer physics were readily developed [13]. The last part of this section is based on [16, 21, 22]. Boundary CFT as described in Sect. 3 appeared in Cardy's articles [3–6], see also [33]. The Temperley-Lieb algebra discussed in Sect. 4 appeared in [32]. This section is based on [17, 18, 23, 28, 30]. Section 5 is based on [6, 12, 17, 31].

References

1. Baxter, R.J.: Exactly Solved Models in Statistical Mechanics. Academic Press, London (1982)

2. Baxter, R.J., Kelland, S.B., Wu, F.Y.: Equivalence of the Potts model or Whitney polynomial with an ice-type model. J. Phys. A, Math. Gen. **9**, 397 (1975)
3. Cardy, J.L.: Conformal invariance and surface critical behaviour. Nucl. Phys. B **240**, 514 (1984)
4. Cardy, J.L.: Operator content of two-dimensional conformally invariant theories. Nucl. Phys. B **270**, 186 (1986)
5. Cardy, J.L.: Effect of boundary conditions on the operator content of two-dimensional conformally invariant theories. Nucl. Phys. B **275**, 200 (1986)
6. Cardy, J.L.: The $O(n)$ model on the annulus. J. Stat. Phys. **125**, 1 (2006)
7. den Nijs, M.: Extended scaling relations for the magnetic critical exponents of the Potts model. Phys. Rev. B **27**, 1674 (1983)
8. den Nijs, M.: Extended scaling relations for the chiral and cubic crossover exponents. J. Phys. A, Math. Gen. **17**, 295 (1984)
9. di Francesco, P., Saleur, H., Zuber, J.B.: Relations between the Coulomb gas picture and conformal invariance of two-dimensional critical models. J. Stat. Phys. **49**, 57 (1987)
10. di Francesco, P., Mathieu, P., Sénéchal, D.: Conformal Field-Theory. Springer, Heidelberg (1997)
11. Domany, E., Mukamel, D., Nienhuis, B., Schwimmer, A.: Duality relations and equivalences for models with $O(n)$ and cubic symmetry. Nucl. Phys. B **190**, 279 (1981)
12. Dubail, J., Jacobsen, J.L., Saleur, H.: Conformal two-boundary loop model on the annulus. Nucl. Phys. B **813**, 430 (2009)
13. Duplantier, B., Saleur, H.: Exact critical properties of two-dimensional dense self-avoiding walks. Nucl. Phys. B **290**, 291 (1987)
14. Fortuin, C.M., Kasteleyn, P.W.: On the random-cluster model. I. Introduction and relation to other models. Physica **57**, 536 (1972)
15. Jacobsen, J.L.: Conformal field theory applied to loop models. In: Guttmann, A.J. (ed.) Polygons, Polyominoes and Polycubes. Lecture Notes in Physics, vol. 775, pp. 347–424. Springer, Heidelberg (2009)
16. Jacobsen, J.L., Kondev, J.: Field theory of compact polymers on the square lattice. Nucl. Phys. B **532**, 635 (1998)
17. Jacobsen, J.L., Saleur, H.: Conformal boundary loop models. Nucl. Phys. B **788**, 137 (2008)
18. Jacobsen, J.L., Saleur, H.: Combinatorial aspects of boundary loop models. J. Stat. Mech. P01021 (2008)
19. José, J.V., Kadanoff, L.P., Kirkpatrick, S., Nelson, D.R.: Renormalization, vortices, and symmetry-breaking perturbations in the two-dimensional planar model. Phys. Rev. B **16**, 1217 (1977)
20. Kasteleyn, P.W., Fortuin, C.M.: Phase transitions in lattice systems with random local properties. J. Phys. Soc. Jpn. **26**(Suppl.), 11 (1969)
21. Kondev, J., Henley, C.L.: Four-coloring model on the square lattice: a critical ground state. Phys. Rev. B **52**, 6628 (1995)
22. Kondev, J.: Liouville field theory of fluctuating loops. Phys. Rev. Lett. **78**, 4320 (1997)
23. Martin, P.P., Saleur, H.: The blob algebra and the periodic Temperley-Lieb algebra. Lett. Math. Phys. **30**, 189 (1994)
24. Nienhuis, B.: Exact critical point and critical exponents of $O(n)$ models in two dimensions. Phys. Rev. Lett. **49**, 1062 (1982)
25. Nienhuis, B.: Critical behaviour of two-dimensional spin models and charge asymmetry in the Coulomb gas. J. Stat. Phys. **34**, 731 (1984)
26. Nienhuis, B.: Coulomb gas formulations of two-dimensional phase transitions. In: Domb, C., Lebowitz, J.L. (eds.) Phase Transitions and Critical Phenomena, vol. 11. Academic Press, London (1987)
27. Nienhuis, B.: Exact methods in low-dimensional statistical physics and quantum computing. In: Jacobsen, J., et al. (eds.) Les Houches Summer School, Session LXXXIX. Oxford University Press, London (2009)
28. Pasquier, V., Saleur, H.: Common structures between finite systems and conformal field theories through quantum groups. Nucl. Phys. B **330**, 523 (1990)

29. Potts, R.B.: Some generalized order-disorder transformations. Math. Proc. Camb. Philos. Soc. **48**, 106 (1952)
30. Richard, J.-F., Jacobsen, J.L.: Character decomposition of Potts model partition functions, I: Cyclic geometry. Nucl. Phys. B **750**, 250 (2006)
31. Saleur, H., Bauer, M.: On some relations between local height probabilities and conformal invariance. Nucl. Phys. B **320**, 591 (1989)
32. Temperley, H.N.V., Lieb, E.H.: Relations between the percolation and colouring problem and other graph-theoretical problems associated with regular planar lattices: some exact results for the percolation problem. Proc. R. Soc. Lond. A **322**, 251 (1971)
33. Yung, C.M., Batchelor, M.T.: O(n) model on the honeycomb lattice via reflection matrices: Surface critical behaviour. Nucl. Phys. B **453**, 552 (1995)

Index